EXPERIMENTS FOR
ELECTRONIC
PRINCIPLES

FIFTH EDITION

A laboratory manual for use with **Electronic Principles**, Fifth Edition

ALBERT PAUL MALVINO, PH.D., E.E.

GLENCOE
McGraw-Hill

New York, New York   Columbus, Ohio   Mission Hills, California   Peoria, Illinois

Cover photographs:
  Special Effects, Charly Franklin/FPG International
  Microchip, Michael Simpson/FPG International
  Handmade Rice Paper, COMSTOCK, Inc./Michael Stuckey

Experiments in Electronic Principles, Fifth Edition

Imprint 1997

Send all inquiries to:
Glencoe/McGraw-Hill
936 Eastwind Drive
Westerville, OH 43081

ISBN 0-02-800847-2

Printed in the United States of America.

6 7 8 9 10 11 066 01 00 99 98 97

CONTENTS

PREFACE

Marvin Minsky, a leading researcher in the field of artificial intelligence, once said, "You don't understand anything until you learn it in more than one way." The concepts that you read about in your textbook need to be learned in a practical way, using real circuits on a laboratory bench. This laboratory manual contains 58 experiments to help you learn the theory that you studied in *Electronic Principles* in another way.

Each experiment is straightforward. It begins with a short theory section that introduces the major ideas in the experiment. The required reading list cites those sections of the textbook that you should read before attempting the experiment. In the procedure section, you will build and test the circuit. Also included are some optional experiments in trouble-shooting, design, and computer programming. The questions at the end of each experiment are a final check on what you have learned during the experiment.

You will find these experiments interesting and at times exciting. They prove, expand, and dramatize the theory that was presented in *Electronic Principles*; they make it come alive. If you will try to learn as much as you can while doing these experiments, you will be that much closer to mastering the art of electronics.

Albert Paul Malvino

EXPERIMENT 1

VOLTAGE AND CURRENT SOURCES

▼

An ideal or perfect voltage source produces an output voltage that is independent of the load resistance. A real voltage source, however, has a small internal resistance that produces an *IR* drop. As long as this internal resistance is much smaller than the load resistance, almost all the source voltage appears across the load. A stiff voltage source is one whose internal resistance is less than 1/100 of the load resistance. With a stiff voltage source, at least 99 percent of the source voltage appears across the load resistor.

A current source is different. It produces an output current that is independent of the load resistance. One way to build a current source is to use a source resistance that is much larger than the load resistance. An ideal current source has an infinite source resistance. A real current source has an extremely high source resistance. A stiff current source is one whose internal resistance is at least 100 times greater than the load resistance. With a stiff current source, at least 99 percent of the source current passes through the load resistor.

In this experiment you will build voltage and current sources, verifying the conditions necessary to get stiff sources. As options, you can also troubleshoot and design sources.

REQUIRED READING

Chapter 1 (Secs. 1-1 and 1-2) of *Electronic Principles*, 5th ed.

EQUIPMENT

1 power supply: adjustable to 10 V
6 ½-W resistors: 10 Ω, 47 Ω, 100 Ω, 470 Ω, 1 kΩ, 10 kΩ
1 VOM (analog or digital multimeter)

PROCEDURE

Voltage Sources

1. The circuit left of the *AB* terminals in Fig. 1-1 represents a voltage source and its internal resistance *R*. Before you measure any voltage or current, you should have an estimate of its value. Otherwise, you really don't know what you're doing. Look at Fig. 1-1 and estimate the load voltage for each value of *R* listed in Table 1-1. Record your rough estimates. Don't use a calculator to get these load voltages. Instead, mentally work out ballpark answers. All you're trying to do here is get into the habit of mentally estimating values before they are measured.

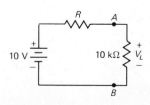

Figure 1-1

2. Connect the circuit in Fig. 1-1 using the values of R given in Table 1-1. Measure and adjust the source voltage to 10 V. For each R value, measure and record V_L.

Current Source

3. The circuit left of the AB terminals in Fig. 1-2 acts like a current source under certain conditions. Estimate and record the load current for each value of load resistance shown in Table 1-2.
4. Connect the circuit of Fig. 1-2 using the R_L values given in Table 1-2. Measure and adjust the source voltage to 10 V. For each R_L value, measure and record I_L.

Troubleshooting (Optional)

5. Connect the circuit of Fig. 1-1 with an R of 470 Ω. Connect a jumper wire between A and B. Measure the voltage across the load resistor and record your answer in Table 1-3.

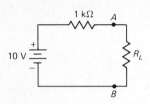

Figure 1-2

6. Remove the jumper and open the load resistor. Measure the load voltage between the AB terminals and record in Table 1-3.

Design (Optional)

7. Select an internal resistance R for the circuit of Fig. 1-1 to get a stiff voltage source for all load resistances greater than 10 kΩ. Connect the circuit of Fig. 1-1 using your design value of R. Measure the load voltage. Record the value of R and the load voltage in Table 1-4.
8. Select an internal resistance R for the circuit of Fig. 1-2 to get a stiff current source for all load resistances less than 100 Ω. Connect the circuit with your design value of R and a load resistance of 100 Ω. Measure the load current. Record the value of R and the load current in Table 1-4.

Computer (Optional)

9. Enter and run this program:

```
10 PRINT "EXPERIMENT 1"
20 PRINT "VOLTAGE AND CURRENT
   SOURCES"
30 END
```

10. Write and run a program with three PRINT statements on lines 10, 20, and 30 that prints your name on the first line, address on the second line, and city, state, and zip code on the third line.

DATA FOR EXPERIMENT 1

Table 1-1. Voltage Source

R	Estimated V_L	Measured V_L
0 Ω		
10 Ω		
100 Ω		
470 Ω		

Table 1-2. Current Source

R_L	Estimated I_L	Measured I_L
0 Ω		
10 Ω		
47 Ω		
100 Ω		

Table 1-3. Troubleshooting

Trouble	Measured V_L
Shorted load	
Open load	

Table 1-4. Design

Type	R	Measured Quantity
Voltage source		
Current source		

QUESTIONS FOR EXPERIMENT 1

1. The data of Table 1-1 prove that load voltage is: ( )
   (a) perfectly constant; (b) small; (c) heavily dependent on load resistance; (d) approximately constant.
2. When internal resistance R increases in Fig. 1-1, load voltage: ( )
   (a) increases slightly; (b) decreases slightly; (c) stays the same.
3. In Fig. 1-1, the voltage source is stiff when R is less than: ( )
   (a) 0 Ω; (b) 100 Ω; (c) 500 Ω; (d) 1 kΩ.
4. The circuit left of the *AB* terminals in Fig. 1-2 acts approximately like a ( )
   current source because the current values in Table 1-2:
   (a) increase slightly; (b) are almost constant; (c) decrease a great deal; (d) depend heavily on load resistance.

5. In Fig. 1-2, the circuit acts like a stiff current source as long as the load ( )
   resistance is:
   (a) less than 10 Ω; (b) large; (c) much larger than 1 kΩ; (d) greater
   than 1 kΩ.

6. Briefly explain the difference between a stiff voltage source and a stiff current source.
   Use the following space:

Troubleshooting (Optional)

7. Explain why the load voltage with a shorted load is zero in Table 1-3. Consider using
   Ohm's law in your explanation.

8. Briefly explain why the load voltage with an open load is approximately equal to the
   source voltage in Table 1-3. Consider using Ohm's law and Kirchhoff's voltage law
   in your explanation.

Design (Optional)

9. You are designing a current source that must appear stiff to all load resistances less
   than 10 kΩ. What is the minimum internal resistance your source can have? Explain
   why you selected this answer:

10. Optional: Instructor's question.

EXPERIMENT 2

THEVENIN'S THEOREM

▼

The Thevenin voltage is the voltage that appears across the load terminals when you open the load resistor. The Thevenin voltage is also called the open-circuit or open-load voltage. The Thevenin resistance is the resistance between the load terminals with the load disconnected and all sources reduced to zero. This means replacing voltage sources by short circuits and current sources by open circuits.

In this experiment you will calculate the Thevenin voltage and resistance of a circuit. Then you will measure these quantities. Also included are troubleshooting and design options.

REQUIRED READING

Chapter 1 (Sec. 1-3) of *Electronic Principles*, 5th ed.

EQUIPMENT

1 power supply: 15 V (adjustable)
7 ½-W resistors: 470 Ω, two 1 kΩ, two 2.2 kΩ, two 4.7 kΩ
1 potentiometer: 5 kΩ
1 VOM (analog or digital multimeter)

PROCEDURE

1. In Fig. 2-1a, calculate the Thevenin voltage V_{TH} and the Thevenin resistance R_{TH}. Record these values in Table 2-1.
2. With the Thevenin values just found, calculate the load voltage V_L across an R_L of 1 kΩ (see Fig. 2-1b). Record V_L in Table 2-2.
3. Also calculate the load voltage V_L for an R_L of 4.7 kΩ as shown in Fig. 2-1c. Record the calculated V_L in Table 2-2.
4. Connect the circuit of Fig. 2-1a, leaving out R_L.
5. Measure and adjust the source voltage to 15 V. Measure V_{TH} and record the value in Table 2-1.
6. Replace the 15-V source by a short circuit. Measure the resistance between the AB terminals using a

convenient resistance range of the VOM. Record R_{TH} in Table 2-1. Now replace the short by the 15-V source.

7. Connect a load resistance R_L of 1 kΩ between the AB terminals of Fig. 2-1a. Measure and record load voltage V_L (Table 2-2).
8. Change the load resistance from 1 kΩ to 4.7 kΩ. Measure and record the new load voltage.
9. Find R_{TH} by the matched-load method; that is, use the potentiometer as a variable resistance between the AB terminals of Fig. 2-1a. Vary resistance until

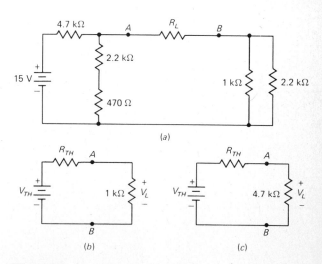

(a)

(b)   *(c)*

Figure 2-1

load voltage drops to half of the measured V_{TH}. Then disconnect the load resistance and measure its resistance with the VOM. This value should agree with R_{TH} found in Step 6.

Troubleshooting (Optional)

10. Put a jumper wire across the 2.2-kΩ resistor, the one on the left side of Fig. 2-1a. Estimate the Thevenin voltage and Thevenin resistance for this trouble and record your rough estimates in Table 2-3. Measure the Thevenin voltage and Thevenin resistance (similar to Steps 5 and 6). Record the measured data in Table 2-3.
11. Remove the jumper wire and open the 2.2-kΩ resistor of Fig. 2-1a. Estimate and record the Thevenin quantities (Table 2-3). Measure and record the Thevenin quantities.

Design (Optional)

12. Select the resistors for the unbalanced Wheatstone bridge of Fig. 2-2 to meet these specifications: $V_{TH} = 4.35$ V and $R_{TH} = 3$ kΩ. Record your design values in Table 2-4. Connect your circuit. Measure and record the Thevenin quantities.

Computer (Optional)

13. Enter and run this program:

```
10 X = 16
20 Y = 27
30 Z = X+Y
40 PRINT Z
50 END
```

14. Write and run a program that prints the total resistance of two resistors in series. One of the resistances is 220 Ω, and the other is 470 Ω.

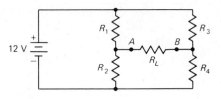

Figure 2-2

DATA FOR EXPERIMENT 2

Table 2-1. Thevenin Values

	V_{TH}	R_{TH}
Calculated		
Measured		

Table 2-2. Load Voltages

	V_L for 1 kΩ	V_L for 4.7 kΩ
Calculated		
Measured		

Table 2-3. Troubleshooting

	Estimated		Measured	
	V_{TH}	R_{TH}	V_{TH}	R_{TH}
Shorted 2.2 kΩ				
Open 2.2 kΩ				

Table 2-4. Design

Design values: $R_1 = $ _____

$R_2 = $ _____

$R_3 = $ _____

$R_4 = $ _____

Measured values: $V_{TH} = $ _____

$R_{TH} = $ _____

QUESTIONS FOR EXPERIMENT 2

1. In this experiment you measured Thevenin voltage with: ( )
   (a) an ohmmeter;   (b) the load disconnected;   (c) the load in the circuit.
2. You first measured R_{TH} with a: ( )
   (a) voltmeter;   (b) load;   (c) shorted source.
3. You also measured R_{TH} by the matched-load method, which involves: ( )
   (a) an open voltage source;   (b) a load that is open;   (c) varying Thevenin resistance until it matches load resistance;   (d) changing load resistance until load voltage drops to $V_{TH}/2$.
4. Discrepancies between calculated and measured values in Table 2-1 may be ( )
   caused by:
   (a) instrument error;   (b) resistor tolerance;   (c) human error;   (d) all the foregoing.

5. If a black box puts out a constant voltage for all load resistances, the Thevenin ( )
resistance of this box approaches:
(a) zero; (b) infinity; (c) load resistance.

6. Ideally, a voltmeter should have infinite resistance. Explain how a voltmeter with an input resistance of 100 kΩ will introduce a small error in Step 5 of the procedure.

Troubleshooting (Optional)

7. Briefly explain why the Thevenin voltage and resistance are both lower when the 2.2-kΩ resistor is shorted.

8. Explain why V_{TH} and R_{TH} are higher when the 2.2-kΩ resistor is open.

Design (Optional)

9. If you were manufacturing automobile batteries, would you try to produce a very low internal resistance or a very high internal resistance? Explain your reasoning.

10. Optional: Instructor's question.

EXPERIMENT 3

TROUBLESHOOTING

▼

An open device always has zero current and unknown voltage. You have to figure out what the voltage is by looking at the rest of the circuit. On the other hand, a shorted device always has the zero voltage and unknown current. You have to figure out what the current is by looking at the rest of the circuit. In this experiment, you will insert troubles into a basic circuit. Then you will calculate and measure the voltage of the circuit.

REQUIRED READING

Chaper 1 (Sec. 1-5) of *Electronic Principles*, 5th ed.

EQUIPMENT

1 power supply: 10 V
4 ½-W resistors: 1 kΩ, 2.2 kΩ, 3.9 kΩ, and 4.7 kΩ
1 VOM (analog or digital)

PROCEDURE

1. Connect the circuit shown in Fig. 3-1.
2. Calculate the voltage between node A and ground. Record the value in Table 3-1 under "Circuit OK."
3. Calculate the voltage between node B and ground. Record the value.
4. Measure the voltages at A and B. Record these values in Table 3-1.
5. Open resistor R_1. Calculate the voltages at nodes A

and B. Record these values in Table 3-1. Next, measure the voltages at nodes A and B. Record the values.
6. Repeat Step 5 for each of the remaining resistors listed in Table 3-1.
7. Short-circuit resistor R_1 by placing a jumper wire across it. Calculate and record the voltages in Table 3-1.
8. Repeat Step 7 for each of the remaining resistors in Table 3-1.

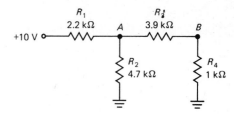

Figure 3-1

DATA FOR EXPERIMENT 3

Table 3-1. Troubles and Voltages

	Calculated		Measured	
Trouble	V_A	V_B	V_A	V_B
Circuit OK				
R_1 open				
R_2 open				
R_3 open				
R_4 open				
R_1 shorted				
R_2 shorted				
R_3 shorted				
R_4 shorted				

QUESTIONS FOR EXPERIMENT 3

1. When R_1 is open in Fig. 3-1, V_A is approximately: (   )
   (a) 0;   (b) 1.06 V;   (c) 1.41 V;   (d) 6.81 V.
2. When R_2 is open in Fig. 3-1, V_B is approximately: (   )
   (a) 0;   (b) 1.06 V;   (c) 1.41 V;   (d) 6.81 V.
3. When R_3 is open in Fig. 3-1, V_A is approximately: (   )
   (a) 0;   (b) 1.06 V;   (c) 1.41 V;   (d) 6.81 V.
4. When R_4 is open in Fig. 3-1, V_B is approximately: (   )
   (a) 0;   (b) 1.06 V;   (c) 1.41 V;   (d) 6.81 V.
5. When R_1 is shorted in Fig. 3-1, V_A is approximately: (   )
   (a) 0;   (b) 2.04 V;   (c) 2.72 V;   (d) 10 V.
6. When R_2 is shorted in Fig. 3-1, V_B is approximately: (   )
   (a) 0;   (b) 2.04 V;   (c) 2.72 V;   (d) 10 V.
7. When R_3 is shorted in Fig. 3-1, V_A is approximately: (   )
   (a) 0;   (b) 2.04 V;   (c) 2.72 V;   (d) 10 V.
8. When R_4 is shorted in Fig. 3-1, V_B is approximately: (   )
   (a) 0;   (b) 1.06 V;   (c) 1.41 V;   (d) 6.81 V.
9. When R_3 is open in Fig. 3-1, V_B is approximately: (   )
   (a) 0;   (b) 1.06 V;   (c) 1.41 V;   (d) 4.92 V.
10. When R_4 is shorted in Fig. 3-1, V_A is approximately: (   )
   (a) 0;   (b) 1.06 V;   (c) 1.41 V;   (d) 4.92 V.

EXPERIMENT 4

THE DIODE CURVE

▼

A resistor is a linear device because its voltage and current are proportional in either direction. A diode, on the other hand, is a nonlinear device because its current and voltage are not proportional. Furthermore, a diode is a unilateral device because it conducts well only in the forward direction. As a guide, a small-signal silicon diode has a dc reverse/forward resistance ratio of more than 1000:1. In this experiment you will measure diode currents and voltages for both forward and reverse bias. This will allow you to draw the diode curve. Also included are troubleshooting, design, and computer operations.

REQUIRED READING

Chapter 2 (Secs. 2-8 to 2-11) and Chap. 3 (Secs. 3-1 to 3-4) of *Electronic Principles*, 5th ed.

EQUIPMENT

1 power supply: adjustable from approximately 0 to 15 V
1 diode: 1N914 (or almost any small-signal silicon diode)
3 ½-W resistors: 220 Ω, 1 kΩ, 100 kΩ
1 VOM (analog or digital multimeter)
1 milliammeter or another VOM if available
Graph paper, rectangular coordinates

PROCEDURE

Ohmmeter Test

1. Using the VOM as an ohmmeter, measure a 1N914's dc forward resistance and reverse resistance on one of the middle resistance ranges. If the diode is all right, you should have a reverse/forward ratio greater than 1000:1.

Diode Data

2. Connect the circuit of Fig. 4-1 using a current-limiting resistor of 1 kΩ. For each source voltage listed in Table 4-1, measure and record the diode voltage V and the diode current I.

3. Calculate and record the dc forward resistance of the diode for each current of Table 4-1.

4. Reverse the source voltage in Fig. 4-1. For each source voltage of Table 4-2, measure and record the diode voltage V and the diode current I.

5. Calculate and record the dc reverse resistance of the diode for each source voltage of Table 4-2.

6. Graph the data of Tables 4-1 and 4-2 to get a diode curve (I versus V).

7. The foregoing steps prove that the diode conducts easily in the forward direction and poorly in the reverse direction. It's like a one-way conductor. With this in mind, estimate the diode current in Fig. 4-2a and b. Record your ballpark estimates in Table 4-3.

8. Connect the circuit of Fig. 4-2a (forward bias). Measure and record the diode current in Table 4-3.

9. Connect the circuit of Fig. 4-2b (reverse bias). Measure and record the diode current.

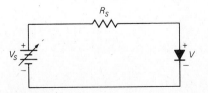

Figure 4-1

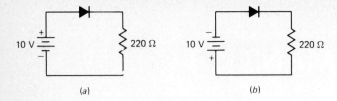

(a)   (b)

Figure 4-2

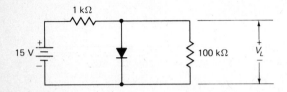

Figure 4-3

Troubleshooting (Optional)

10. Connect the circuit of Fig. 4-3. Estimate the load voltage V_L and record in Table 4-4. Then measure and record V_L.

11. Short the diode with a jumper wire. Estimate V_L for this condition and record in Table 4-4. Measure and record V_L.

12. Remove the jumper wire. Disconnect one end of the diode. Estimate V_L and record. Next, measure and record V_L.

Design (Optional)

13. Select a source voltage and a current-limiting resistance to produce 10 mA in Fig. 4-1a. (Use the resistors from earlier parts of the experiment.) Connect your circuit and measure the current. Record your values of V_S and R_S, along with the measured I in Table 4-5.

Computer (Optional)

14. In BASIC the arithmetic operators are +, −, *, and /. They stand for plus, minus, times, and divide by. Enter and run this program:

```
10 R1 = 4700
20 R2 = 6800
30 RT = R1 + R2
40 PRINT RT
```

15. Write and run a program that calculates and prints the total resistance of 4.7 kΩ in parallel with 6.8 kΩ.

● DATA FOR EXPERIMENT 4

Table 4-1. Forward Bias

V_S	V	I	R
0			(no entry)
0.5 V			
1 V			
2 V			
4 V			
6 V			
8 V			
10 V			
15 V			

Table 4-2. Reverse Bias

V_S	V	I	R
−1 V			
−5 V			
−10 V			
−15 V			

Table 4-3. Diode Conduction

	Estimated I	Measured I
Fig. 4-2a		
Fig. 4-2b		

Table 4-4. Troubleshooting

	Estimated V_L	Measured V_L
Normal diode		
Shorted diode		
Open diode		

Table 4-5. Design

$V_S =$ _____ $R_S =$ _____ $I =$ _____

QUESTIONS FOR EXPERIMENT 4

1. In this experiment the knee or offset voltage is closest to: ( )
   (a) 0.3 V;   (b) 0.7 V;   (c) 1. V;   (d) 1.2 V.
2. With forward bias, the dc resistance decreases when: ( )
   (a) current increases;   (b) diode decreases;   (c) the ratio V/I increases;
   (d) the ratio I/V decreases.
3. A diode acts like a high resistance when: ( )
   (a) its current is large;   (b) forward-biased;   (c) reverse-biased;
   (d) shorted.
4. Which of the following approximately describes the diode curve above the ( )
   forward knee?
   (a) it becomes horizontal;   (b) voltage increases rapidly;   (c) current increases rapidly;   (d) dc resistance increases rapidly.
5. Which of the following describes the diode curve in the reverse direction? ( )
   (a) ratio I/V is high;   (b) it becomes vertical below breakdown;   (c) dc resistance is low;   (d) current is approximately zero below breakdown.
6. Briefly describe how a diode differs from an ordinary resistor:

Troubleshooting (Optional)

7. Why is the load voltage around 0.7 V in Fig. 4-3 when the diode is okay?

8. Why is the load voltage slightly less than 15 V when the diode is open in Fig. 4-3?

Design (Optional)

9. If you are trying to set up a fixed current through any diode, is it better to use a low or a high source voltage? Explain your reasoning.

10. Optional: Instructor's question.

EXPERIMENT 5

DIODE APPROXIMATIONS

▼

In the ideal or first approximation, a diode acts like a closed switch when forward-biased and an open switch when reversed-biased. In the second approximation, we include the knee voltage of the diode when it is forward-biased. This means assuming 0.7 V across a conducting silicon diode (0.3 V for germanium). The third approximation includes the knee voltage and the bulk resistance; because of this, the voltage across a conducting diode increases as the diode current increases. In troubleshooting and design, the second approximation is usually adequate.

In this experiment, you will work with the three diode approximations. Also included are troubleshooting, design, and computer options.

REQUIRED READING

Chapter 3 (Secs. 3-5 to 3-9) of *Electronic Principles*, 5th ed.

EQUIPMENT

1 power supply: adjustable from approximately 0 to 15 V
1 diode: 1N914 (or almost any small-signal silicon diode)
3 ½-W resistors: two 220 Ω, 470 Ω
1 VOM (analog or digital multimeter)

PROCEDURE

1. Connect the circuit of Fig. 5-1a. Adjust the source to set up a current of 10 mA through the diode. Estimate the diode voltage V and record in Table 5-1.

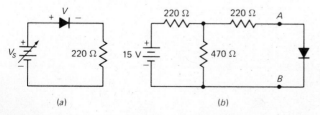

Figure 5-1

2. Measure the diode voltage V and record in Table 5-1.
3. Adjust the source to get 50 mA. Estimate the diode voltage and record in Table 5-1. Measure and record diode voltage V.
4. In this experiment, we will let the knee voltage be the measured diode voltage for a diode current of 10 mA. Record the knee voltage in Table 5-2. (It should be in the vicinity of 0.7 V.)
5. Calculate the bulk resistance using

$$r_B = \frac{\Delta V}{\Delta I}$$

where ΔV and ΔI are the changes in measured voltage and current in Table 5-1. Record r_B in Table 5-2.
6. Calculate the diode current in Fig. 5-1b as follows: thevenize the circuit left of the AB terminals. Then calculate the diode current with the ideal, second, and third approximations (use the V_{knee} and r_B of Table 5-2). Record your answers in Table 5-3.
7. Connect the circuit of Fig. 5-1b. Measure and record the diode current (Table 5-3).

Troubleshooting (Optional)

8. Estimate the diode current in Fig. 5-1b for each of these conditions: 470 Ω shorted and open. Record your rough estimates in Table 5-4.

9. Measure and record the diode current in the circuit of Fig. 5-1*b* with the 470-Ω resistor shorted and open.

Design (Optional)

10. Using the second approximation in Fig. 5-2, select values for resistors and source voltage to produce a diode current of approximately 8.9 mA. (Use the same resistance values as Fig. 5-1*b*, although you can move the values.) Connect your design and mea-

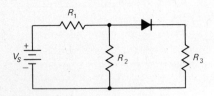

Figure 5-2

sure the diode current. Record all data listed in Table 5-5.

Computer (Optional)

11. Enter and run this program:

```
10 PRINT "ENTER R1"
20 INPUT R1
30 PRINT "R1 EQUALS"
40 PRINT R1
50 END
```

12. Write and run a program that calculates and prints out the Thevenin voltage, Thevenin resistance, and diode current in Fig. 5-1*b*. Use the INPUT statements to enter the data and use the second approximation of a diode.

DATA FOR EXPERIMENT 5

Table 5-1. Two Points on the Forward Curve

I	Estimated V	Measured V
10 mA		
50 mA		

Table 5-2. Diode Values

$V_{knee} =$

$r_B =$

Table 5-3. Diode Current

Ideal $I =$

Second $I =$

Third $I =$

Measured $I =$

Table 5-4. Troubleshooting

	Estimated I	Measured I
Shorted 470 Ω		
Open 470 Ω		

Table 5-5. Design

	Design 1	Design 2	Design 3
$V_S =$			
$R_1 =$			
$R_2 =$			
$R_3 =$			
I (diode) $=$			

QUESTIONS FOR EXPERIMENT 5

1. In this experiment, knee voltage is the diode voltage that: (   )
   (a) equals 0.3 V; (b) equals 0.7 V; (c) corresponds to 10 mA;
   (d) corresponds to 50 mA.

2. Bulk resistance is: ( )
   (a) diode voltage divided by current; (b) the ratio of voltage difference to current difference above the knee; (c) the same as the dc resistance of the diode; (d) none of the foregoing.
3. The dc resistance of a silicon diode for a current of 10 mA is closest to: ( )
   (a) 2.5 Ω; (b) 10 Ω; (c) 70 Ω; (d) 1 kΩ.
4. In Fig. 5-1b, the power dissipated by the diode is closest to: ( )
   (a) 0; (b) 1.5 mW; (c) 15 mW; (d) 150 mW.
5. Suppose the diode of Fig. 5-1b has an $I_{F(\max)}$ of 500 mA. To avoid diode ( ) damage, the source voltage can be no more than
   (a) 15 V; (b) 50 V; (c) 185 V; (d) 272 V.
6. The steeper the diode curve, the smaller the bulk resistance. Explain why this is true.

Troubleshooting (Optional)

7. Explain why there is no diode current when the 470-Ω resistor is shorted in Fig. 5-1b.

8. Why does the diode current increase with an open 470-Ω resistor in Fig. 5-1b?

Design (Optional)

9. How many designs are possible in Step 10 of the Procedure? ( )
   (a) 1; (b) 2; (c) 3; (d) 4.
10. Optional: Instructor's question.

EXPERIMENT 6

RECTIFIER CIRCUITS

▼

The three basic rectifier circuits are the half-wave, the full-wave, and the bridge. The ripple frequency of a half-wave rectifier is equal to the input frequency, whereas the ripple frequency of a full-wave or bridge rectifier is equal to twice the input frequency. For a given transformer, the unfiltered output of the half-wave and full-wave rectifiers ideally has a dc value of slightly less than half the rms secondary voltage (45 percent), while the unfiltered output of a bridge rectifier is slightly less than the rms secondary voltage (90 percent).

In this experiment you will build all three types of rectifiers and measure their input-output characteristics. Be especially careful in this experiment when connecting the transformer to line voltage. The transformer should have a fused line cord with all primary connections insulated to avoid electrical shock.

REQUIRED READING

Chapter 4 (Secs. 4-1 to 4-4) of *Electronic Principles*, 5th ed.

EQUIPMENT

1 transformer, 12.6 V ac center-tapped (Triad F-25X or equivalent) with fused line cord
4 silicon diodes: 1N4001 (or equivalent)
1 ½-W resistor: 1 kΩ
1 VOM (analog or digital multimeter)
1 oscilloscope

PROCEDURE

Half-Wave Rectifier

1. In Fig. 6-1a, the rms secondary output voltage is a nominal 12.6 V ac. Calculate the peak output voltage across the 1-kΩ load resistor. Also calculate the dc output voltage and ripple frequency. Record your calculations in Table 6-1.
2. Connect the half-wave rectifier shown in Fig. 6-1a.
3. Measure the rms voltage across the secondary winding and record in Table 6-1.

4. Measure and record the dc load voltage.
5. Use an oscilloscope to look at the rectified voltage across the 1-kΩ load resistor. Record the peak voltage of the half-wave signal. Next, measure the period of the rectified output. Calculate the ripple frequency and record the result in Table 6-1.

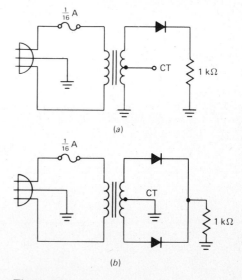

Figure 6-1

Full-Wave Rectifier

6. In Fig. 6-1*b*, calculate and record the quantities listed in Table 6-2.
7. Connect the center-tap rectifier of Fig. 6-1*b*.
8. Measure and record the quantities listed in Table 6-2.

Bridge Rectifier

9. In Fig. 6-2, calculate the quantities listed in Table 6-3.
10. Connect the bridge rectifier of Fig. 6-2.
11. Measure and record the quantities listed in Table 6-3.

Troubleshooting (Optional)

12. Assume one of the diodes is open in the bridge rectifier. Calculate and record the dc output voltage and ripple frequency in Table 6-4.
13. Open one of the diodes. Measure and record the dc output voltage and ripple frequency. Restore the diode to a normal connection.
14. Assume half of the secondary winding to the bridge rectifier is shorted (between the center tap and either end). Calculate and record the dc output voltage and ripple frequency in Table 6-4.
15. Simulate the foregoing short by disconnecting either end of the secondary and connecting the center tap in its place. Measure and record the dc output voltage and ripple frequency.

Design (Optional)

16. Figure out how to modify the bridge rectifier of Fig. 6-2 to meet the following specifications: dc load

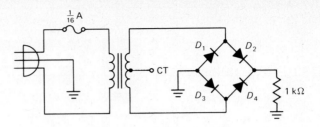

Figure 6-2

voltage is approximately 5.67 V, and dc load current is approximately 20 mA. (You need to select a new load resistor.)

17. Get the required load resistor and connect the modified circuit. Measure and record all the quantities listed in Table 6-5.

Computer (Optional)

18. Enter and run this program:

```
10 PRINT "ENTER RMS SECONDARY
   VOLTAGE"
20 INPUT V2
30 PRINT "DC OUTPUT VOLTAGE IS"
40 PRINT 0.9 * V2
```

19. Write and run a program that inputs the rms secondary voltage to a bridge rectifier and prints out the peak output voltage, dc output voltage, dc diode current, peak inverse voltage, and ripple frequency.

DATA FOR EXPERIMENT 6

Table 6-1. Half-Wave Rectifier

	Calculated	Measured
RMS secondary voltage	12.6 V	
Peak output voltage		
DC output voltage		
Ripple frequency		

Table 6-2. Full-Wave Rectifier

	Calculated	Measured
RMS secondary voltage	12.6 V	
Peak output voltage		
DC output voltage		
Ripple frequency		

Table 6-3. Bridge Rectifier

	Calculated	Measured
RMS secondary voltage	12.6 V	
Peak output voltage		
DC output voltage		
Ripple frequency		

Table 6-4. Troubleshooting

	Calculated		Measured	
	V_{dc}	f_{out}	V_{dc}	f_{out}
Diode open				
Half-secondary short				

Table 6-5. Design

	Calculated	Measured
RMS secondary voltage	6.3 V	
Peak output voltage		
DC load voltage		

Table 6-5. (*Continued*)

	Calculated	Measured
DC load current		
Ripple frequency		
Load resistance		

QUESTIONS FOR EXPERIMENT 6

1. To measure the rms secondary voltage, it is best to use: ( )
   (a) an oscilloscope: (b) an ammeter; (c) a voltmeter with the common lead grounded; (d) a floating VOM.
2. With the full-wave rectifier of this experiment, the dc load voltage was closest to: ( )
   (a) 1 V; (b) 3 V; (c) 6 V; (d) 12 V.
3. The dc load voltage out of the bridge rectifier compared with the full-wave rectifier was approximately: ( )
   (a) half as large; (b) the same; (c) twice as large; (d) 60 Hz.
4. Of the three rectifiers tested, the one with the largest dc output was: ( )
   (a) half-wave; (b) full-wave (c) bridge; (d) no answer.
5. The unfiltered dc output voltage from a bridge rectifier is ideally what percent of the rms secondary voltage: ( )
   (a) 31.8; (b) 45; (c) 63.6; (d) 90.
6. Explain why the bridge rectifier is the most widely used of the three types.

Troubleshooting (Optional)

7. Explain why the dc output voltage and ripple frequency of a bridge rectifier drop in half when any diode opens.

8. If any diode in a bridge rectifier is shorted for any reason (solder bridge, fused diode, etc.), what will happen to the other diodes when power is applied? Explain your answer briefly.

Design (Optional)

9. Briefly explain what you did in your design and why you did it.

10. Optional: Instructor's question

EXPERIMENT 7

THE CAPACITOR-INPUT FILTER

▼

By connecting the output of a bridge rectifier to a capacitor-input filter, we can produce a dc load voltage that is approximately constant. Ideally, the filtered dc output voltage equals the peak secondary voltage. To a better approximation, the dc voltage is typically 90 to 95 percent of the peak secondary voltage with a peak-to-peak ripple of about 10 percent.

In this experiment you will connect a bridge rectifier to a capacitor-input filter. By changing load resistors and filter capacitors, you will verify the basic relations discussed in the textbook. Be especially careful in this experiment when connecting the transformer to line voltage. The transformer should have a fused line cord with all primary connections insulated to avoid electrical shock.

REQUIRED READING

Chapter 4 (Secs. 4-5 to 4-8) of *Electronic Principles*, 5th ed.

EQUIPMENT

1 transformer: 12.6 V ac center-tapped (Triad F-25X or equivalent) with fused line cord
4 silicon diodes: 1N4001 (or equivalent)
2 ½-W resistors: 1 kΩ, 10 kΩ
2 capacitors: 47 µF and 470 µF (25-V rating or better)
1 VOM (analog or digital multimeter)
1 oscilloscope

PROCEDURE

1. Measure the resistance of the primary and secondary windings. Record in Table 7-1.
2. In Fig. 7-1, assume the rms secondary voltage is 12.6 V. Also assume $R_L = 1$ kΩ and $C = 47$ µF. Calculate and record the quantities listed in Table 7-2.
3. Build the circuit of Fig. 7-1 with $R_L = 1$ kΩ and $C = 47$ µF.
4. Measure and record all the quantities listed in Table 7-2.
5. Repeat Steps 2 through 4 for $R_L = 1$ kΩ and $C = 470$ µF. Use Table 7-3.

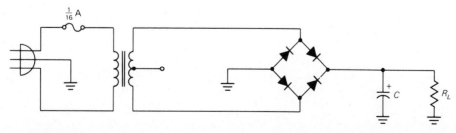

Figure 7-1

6. Repeat Steps 2 through 4 for $R_L = 10$ kΩ and $C = 470$ μF. Use Table 7-4.

Troubleshooting (Optional)

7. Assume one of the diodes is open in Fig. 7-1 with $R_L = 1$ kΩ and $C = 470$ μF. Calculate the dc load voltage, ripple frequency, and peak-to-peak ripple. Record your results in Table 7-5.
8. Connect the foregoing circuit with one of the diodes open. Measure and record the quantities of Table 7-5.
9. Assume the filter capacitor is open in Fig. 7-1 with $R_L = 1$ kΩ and $C = 470$ μF. Calculate and record the quantities listed in Table 7-5 for this trouble.
10. Connect the circuit of Fig. 7-1 with an open filter capacitor. Measure and record the remaining quantities of Table 7-5.

Design (Optional)

11. Select a filter capacitor for the circuit of Fig. 7-1 to get a peak-to-peak ripple of about 10 percent of load voltage for an R_L of 3.9 kΩ. Calculate and record the quantities of Table 7-6.
12. Connect your circuit. Measure and record the quantities of Table 7-6.

Computer (Optional)

13. Write and run a program that calculates and prints out the peak-to-peak ripple of a bridge rectifier driving a capacitor-input filter. The inputs are dc load current and capacitance. Assume a line frequency of 60 Hz.

● DATA FOR EXPERIMENT 7

Table 7-1. Transformer Resistances _____

$R_{pri} =$ _____

$R_{sec} =$ _____

Table 7-2. $R_L = 1\ k\Omega$ and $C = 47\ \mu F$

	Calculated	Measured
RMS secondary voltage	12.6 V	
Peak output voltage		
DC output voltage		
DC load current		
Ripple frequency		
Peak-to-peak ripple		

Table 7-3. $R_L = 1\ k\Omega$ and $C = 470\ \mu F$

	Calculated	Measured
RMS secondary voltage	12.6 V	
Peak output voltage		
DC output voltage		
DC load current		
Ripple frequency		
Peak-to-peak ripple		

Table 7-4. $R_L = 10\ k\Omega$ and $C = 470\ \mu F$

	Calculated	Measured
RMS secondary voltage	12.6 V	
Peak output voltage		
DC output voltage		
DC load current		
Ripple frequency		
Peak-to-peak ripple		

Table 7-5. Troubleshooting

	Calculated			Measured		
	V_{dc}	f_{out}	V_{rip}	V_{dc}	f_{out}	V_{rip}
Open diode						
Open capacitor						

Table 7-6. Design for 10-Percent Ripple

	Calculated	Measured
RMS secondary voltage		
Peak output voltage		
DC load voltage		
DC load current		
Ripple frequency		
Peak-to-peak ripple		

QUESTIONS FOR EXPERIMENT 7

1. In this experiment the dc output voltage from the capacitor-input filter was ( )
   approximately equal to:
   (a) peak primary voltage; (b) peak secondary voltage; (c) rms primary
   voltage; (d) rms secondary voltage.

2. The peak-to-peak ripple decreases when the: ( )
   (a) load resistance decreases; (b) filter capacitor decreases; (c) ripple frequency decreases; (d) filter capacitor increases.

3. The turns ratio of the transformer is approximately 9:1. Because of the ( )
   transformer resistances in Table 7-1, the Thevenin resistance facing the filter
   capacitor must be at least:
   (a) $R_{sec} + R_{pri}/3$; (b) $R_{sec} + R_{pri}/9$; (c) $R_{sec}/9 + R_{pri}$; (d) $R_{sec} + R_{pri}/81$.

4. For normal operation, the ripple frequency is: ( )
   (a) 0; (b) 60 Hz; (c) 120 Hz; (d) 240 Hz.

5. When the load resistance increases, the peak-to-peak ripple: ( )
   (a) decreases; (b) stays the same; (c) increases; (d) none of the foregoing.

6. Briefly explain how a capacitor-input filter works.

Troubleshooting (Optional)

7. When any diode opens, the circuit of Fig. 7-1 becomes a capacitor-input filter ( )
   driven by a:
   (a) half-wave rectifier; (b) full-wave rectifier; (c) bridge rectifier;
   (d) unilateral converter.

8. Briefly explain what happens to the circuit of Fig. 7-1 when the filter capacitor opens.

Design (Optional)

9. What size capacitor did you use in your design? Why did you select this size?

10. Optional: Instructor's question.

EXPERIMENT 8

VOLTAGE DOUBLERS

▼

A voltage multiplier produces a dc voltage equal to a multiple of the peak input voltage. Voltage multipliers are useful with high voltage/low current loads. With a voltage doubler, you get twice as much dc output voltage as you do from a standard peak rectifier. This is useful when you are trying to produce high voltages (several hundred volts or more) because higher secondary voltages result in bulkier transformers. At some point, a designer may prefer to use voltage doublers instead of bigger transformers. With a voltage tripler, the dc voltage is approximately three times the peak input voltage. As the multiple increases, the peak-to-peak ripple gets worse.

In this experiment, you will build half-wave and full-wave voltage doublers. You will measure the dc output voltage and peak-to-peak ripple of these circuits to verify the operation described in the textbook.

REQUIRED READING

Chapter 4 (Sec. 4-15) of *Electronic Principles*, 5th ed.

EQUIPMENT

1 transformer: 12.6 V ac center-tapped (Triad F-25X or equivalent) with fused line cord
2 silicon diodes: 1N4001 (or equivalent)
1 ½-W resistor: 1 kΩ
2 capacitors: 470 μF (25-V rating or better)
1 VOM (analog or digital multimeter)
1 oscilloscope

PROCEDURE

Half Wave Doubler

1. Measure the resistance of the primary and secondary windings. Record in Table 8-1.
2. In Fig. 8-1, assume the rms secondary voltage is 12.6 V. Calculate and record the quantities listed in Table 8-2. Use Eq. (4-8) in the textbook to calculate the peak-to-peak ripple.
3. Connect the circuit.
4. Measure and record all the quantities listed in Table 8-2.

Full-Wave Doubler

5. Repeat Steps 2 through 4 for the full-wave doubler of Fig. 8-2. Use Table 8-3 to record your data. When calculating the peak-to-peak ripple, notice that the load resistor is in parallel with two capacitors in series.

Troubleshooting (Optional)

6. Assume capacitor C_1 is open in Fig. 8-1.
7. Estimate the dc load voltage, ripple frequency, and peak-to-peak ripple. Record your rough estimates in Table 8-4.
8. Connect the circuit with the foregoing trouble. Measure and record the quantities of Table 8-4.

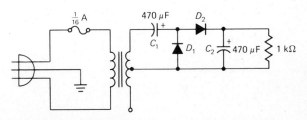

Figure 8-1

9. Assume diode D_2 is open in Fig. 8-1. Repeat Steps 7 and 8.

Design (Optional)

11. Select a filter capacitor (nearest standard size) for the circuit of Fig. 8-1 to get a peak-to-peak ripple of approximately 10 percent of load voltage for an R_L of 3.9 kΩ. Calculate and record the quantities of Table 8-5. Record your design value for capacitance here:

$$C =$$

12. Connect your circuit. Measure and record the quantities of Table 8-5.

Computer (Optional)

13. Write and run a program that calculates the peak-to-peak ripple for the full-wave doubler of Fig. 8-2. The inputs are load voltage, load resistance, and capacitance. Assume a line frequency of 60 Hz.

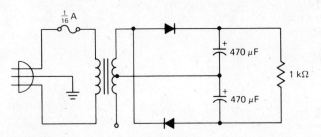

Figure 8-2

DATA FOR EXPERIMENT 8

Table 8-1. Transformer Resistances

R_{pri} = _____

R_{sec} = _____

Table 8-2. Half-Wave Doubler

	Calculated	Measured
Half rms secondary voltage		
DC output voltage		
Ripple frequency		
Peak-to-peak ripple		

Table 8-3. Full-Wave Doubler

	Calculated	Measured
Half rms secondary voltage		
DC output voltage		
Ripple frequency		
Peak-to-peak ripple		

Table 8-4. Troubleshooting

	Estimated			Measured		
	V_{dc}	f_{out}	V_{rip}	V_{dc}	f_{out}	V_{rip}
Open C_1						
Open D_2						
Open C_2						

Table 8-5. Design

	Calculated	Measured
Half rms secondary voltage		
DC load voltage		
Ripple frequency		
Peak-to-peak ripple		

QUESTIONS FOR EXPERIMENT 8

1. In this experiment the dc output voltage from the half-wave doubler was   (   )
   approximately equal to:
   (a) peak primary voltage;   (b) rms secondary voltage;   (c) double the peak
   secondary voltage;   (d) double the peak voltage driving the half-wave doubler.
2. The ripple frequency of a half-wave doubler was: (   )
   (a) 60 Hz;   (b) 120 Hz;   (c) 240 Hz;   (d) 480 Hz.
3. The full-wave doubler has a ripple frequency of: (   )
   (a) 60 Hz;   (b) 120 Hz;   (c) 240 Hz;   (d) 480 Hz.
4. The peak-to-peak ripple of a full-wave doubler compared with a half-wave   (   )
   doubler is:
   (a) half;   (b) the same;   (c) twice as much.
5. Assume the primary resistance is 30 Ω and the secondary resistance is 1 Ω in   (   )
   Fig. 8-2. The primary voltage is 115 V and the secondary voltage is 12.6 V.
   The Thevenin resistance facing either filter capacitor is approximately:
   (a) 0.59 Ω;   (b) 0.86 Ω;   (c) 1.09 Ω;   (d) 1.36 Ω.
6. Briefly explain how the full-wave doubler of Fig. 8-2 works.

Troubleshooting (Optional)

7. Explain why the peak-to-peak ripple is so large for an open C_2 in Table 8-4.

8. Suppose either filter capacitor in Fig. 8-2 is shorted. Explain what happens to the
   nearest diode.

Design (Optional)

9. Justify your design; that is, why did you use the filter capacitor you selected?

10. Optional: Instructor's question.

EXPERIMENT 9

LIMITERS AND PEAK DETECTORS

▼

A positive limiter clips off positive parts of the input signal, and a negative limiter clips negative parts. In a biased limiter, the clipping level is selectable. With a combination limiter, positive and negative parts of the signal are removed. A diode clamp is an alternative name for a limiter. Often, a diode clamp is used to protect a load from excessively high input voltages.

In this experiment, you will connect different limiters. You will also experiment with a peak detector, a variation of the rectifier circuits discussed earlier. A peak detector produces a dc output voltage approximately equal to the peak voltage of the input signal.

REQUIRED READING

Chapter 4 (Sec. 4-16) of *Electronic Principles*, 5th ed.

EQUIPMENT

- 1 audio generator
- 1 power supply: adjustable from approximately 0 to 15 V
- 2 diodes: 1N914 (or almost any small-signal silicon diode)
- 4 ½-W resistors: 470 Ω, 1 kΩ, 10 kΩ, 100 kΩ
- 1 capacitor: 1 μF (10-V rating or better)
- 1 VOM (analog or digital multimeter)
- 1 oscilloscope

PROCEDURE

Positive Limiter

1. In Fig. 9-1, estimate the positive and negative peak output voltages. Record in Table 9-1.
2. Connect the positive limiter of Fig. 9-1. (The 1 kΩ is a dc return in case the source is capacitively coupled.) Adjust the source to get 1 kHz and 20 V peak-to-peak across the input (equivalent to a peak input of 10 V).
3. Move the oscilloscope leads to output. You should get a positively clipped sine wave. Record the pos-

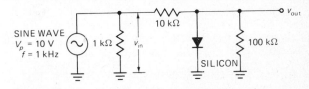

Figure 9-1

itive and negative peak values in Table 9-1. (You must use the dc input of the oscilloscope.)

Negative Limiter

4. In Fig. 9-1, assume the diode polarity is reversed. Record your estimates of the positive and negative output peak voltages in Table 9-1. Reverse the polarity of the diode in your built-up circuit and look at the output waveform. It should be negatively clipped. Record the positive and negative peak values.

Combination Limiter

5. In Fig. 9-2, estimate the positive and negative peak output voltages. Record your estimates in Table 9-1. Connect the combination limiter.
6. Look at the output waveform. Measure and record the positive and negative peaks.

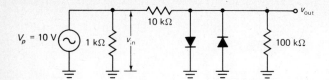

Figure 9-2

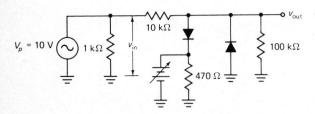

Figure 9-3

Biased Limiter

7. In Fig. 9-3, estimate the output peak voltages and record in Table 9-1. Connect the variable limiter of Fig. 9-3.

8. Look at the output with an oscilloscope (dc input). When you vary the dc source, the positive clipping level should vary from a low value to a high value. If it does, write "variable" under positive peak in Table 9-1. Measure and record the negative peak.

Peak Detector

9. In Fig. 9-4, estimate the dc output voltage, ripple frequency, and peak-to-peak ripple. You may use Eq. (4-8) in the textbook for the latter. Record your estimates in Table 9-2.

10. Connect the peak detector of Fig. 9-4. Adjust the source to get 1 kHz and 10 V peak across the input.

11. Look at the output voltage with the oscilloscope. It should be a dc voltage with an extremely small ripple.

12. Use the VOM to measure the dc output voltage. Record this as V_{dc}.

13. Switch to ac input on the oscilloscope and increase sensitivity until you can measure the ripple accurately. Record the ripple frequency and peak-to-peak ripple.

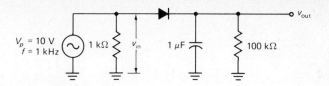

Figure 9-4

14. Because the VOM has input resistance on its voltmeter ranges, it may change the resistance across the 1-μF capacitor. While looking at the output ripple on the oscilloscope, connect and disconnect the VOM. What happens to the ripple when the VOM is disconnected? Record "bigger," "same," or "smaller" in Table 9-2.

Troubleshooting (Optional)

15. In Fig. 9-2, assume the left diode is open. Estimate the positive and negative output peak voltages. Record your estimates in Table 9-3.

16. Connect the circuit of Fig. 9-2 with the left diode open. Measure and record the output peak voltages.

17. Repeat Steps 15 and 16 for a shorted diode.

Design (Optional)

18. Assume the peak voltage is 10 V and the frequency is 5 kHz in Fig. 9-4. Select a filter capacitor (nearest standard size) that produces a peak-to-peak output ripple of approximately 0.5 V. Calculate and record all quantities listed in Table 9-4.

19. Connect the circuit with the new filter capacitor. Adjust the source voltage to 20 V peak-to-peak and the frequency to 5 kHz. Measure and record all quantities listed in Table 9-4.

Computer (Optional)

20. Write and run a program that calculates and prints out the peak-to-peak ripple of a peak detector. The inputs are peak input voltage, frequency, capacitance, and load resistance.

DATA FOR EXPERIMENT 9

Table 9-1. Limiters

| | Estimated | | Measured | |
	Pos Peak	Neg Peak	Pos Peak	Neg Peak
Positive limiter				
Negative limiter				
Combination limiter				
Biased limiter				

Table 9-2. Peak Detector

	Estimated	Measured
V_{dc}		
f_{out}		
V_{rip}		
Ripple change	(no entry)	

Table 9-3. Troubleshooting

| | Estimated | | Measured | |
	Pos Peak	Neg Peak	Pos Peak	Neg Peak
Open diode				
Shorted diode				

Table 9-4. Design

	Calculated	Measured
Capacitance		(no entry)
DC output voltage		
Ripple frequency		
Peak-to-peak ripple		

QUESTIONS FOR EXPERIMENT 9

1. In a negative limiter, which of these is the largest?   (   )
   (a) positive peak;   (b) negative peak;   (c) knee voltage;   (d) crossover
   voltage.
2. The combination limiter of Fig. 9-2:   (   )
   (a) puts out a small sine wave;   (b) generates a small squarish wave;
   (c) has an adjustable clipping level;   (d) has an output proportional to the
   input.

3. When the dc source of Fig. 9-3 varies from 0 to 15 V, the positive output ( )
peak varies from roughly:
(a) 0 to $V_P/2$; (b) 0 to V_P; (c) 0 to $2V_P$; (d) 0 to 0.7 V.

4. In the combination limiter of Fig. 9-2, which diode approximation is the most ( )
reasonable compromise?
(a) ideal; (b) second; (c) third; (d) fourth.

5. The peak-to-peak ripple out of the peak detector of Fig. 9-4 was approxi- ( )
mately what percent of the dc output voltage?
(a) 1%; (b) 5%; (c) 10%; (d) 20%.

6. Briefly explain the operation of the biased combination clipper (Fig. 9-3).

Troubleshooting (Optional)

7. Explain why each trouble in Table 9-3 produces the recorded outputs.

8. You are troubleshooting a peak detector like Fig. 9-4. If the output is a half-wave
rectified sine wave, what is the trouble?

Design (Optional)

9. Which diode approximation appears to be the best compromise for designing peak
detectors? Explain your reasoning.

10. Optional: Instructor's question.

DC CLAMPERS AND PEAK-TO-PEAK DETECTORS

▼

In a dc clamper, a capacitor is charged to approximately the peak input voltage V_P. Depending on the polarity of the charge, the output voltage has a dc component equal to the positive or negative peak input voltage. The output of a positive clamper ideally swings from 0 to $+2V_P$, while the output of a negative clamper swings from 0 to $-2V_P$.

A peak-to-peak detector is a cascaded connection of a dc clamper and a peak detector. The dc clamper ideally produces an output that swings from 0 to $2V_P$, and the peak detector produces a dc output of approximately $2V_P$. Since the final dc output equals the peak-to-peak input voltage, the overall circuit is called a peak-to-peak detector.

If a signal source is capacitively coupled, the problem of the dc return may arise with diode and transistor circuits. When the source has to supply more current on one half-cycle than the other, its coupling capacitor will charge to approximately the peak of the source voltage. Because of this, you will get unwanted dc clamping of the source signal. To eliminate this unwanted clamping, you can add a dc return. It discharges the coupling capacitor and prevents a dc shift of the output signal.

REQUIRED READING

Chapter 4 (Secs. 4-17 to 4-19) of *Electronic Principles*, 5th ed.

EQUIPMENT

1 audio generator
2 diodes: 1N914 (or almost any small-signal silicon diode)
4 ½-W resistors: 1 kΩ, 10 kΩ, 47 kΩ, 100 kΩ
2 capacitors: 1 μF (20-V rating or better)
1 VOM (analog or digital multimeter)
1 oscilloscope

PROCEDURE

Positive Clamper

1. In Fig. 10-1, estimate the positive and negative peaks of the output voltage. Record in Table 10-1.

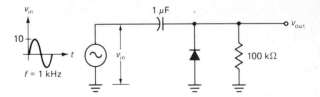

Figure 10-1

2. Connect the positive clamper of Fig. 10-1. Adjust the source to get 1 kHz and 20 V peak-to-peak across the input.
3. With the oscilloscope on dc input, look at the output. It should be a positively clamped sine wave. Measure and record the positive and negative peaks in Table 10-1.
4. Keep the oscilloscope on the output and vary the input voltage. Notice how the negative peak is clamped near zero while the positive peak moves up and down.

Negative Clamper

5. Assume the polarity of the diode in Fig. 10-1 is reversed. Estimate and record the output peaks in Table 10-1.
6. Reverse the polarity of the diode in the built-up circuit. Measure and record the output peaks.

Peak-to-Peak Detector

7. Estimate the dc output voltage and peak-to-peak ripple in Fig. 10-2. You may use Eq. (4-8) for the latter. Record in Table 10-2.
8. Connect the peak-to-peak detector of Fig. 10-2. Adjust the source to get 1 kHz and 20 V peak-to-peak across the input.
9. Look at the voltage across the first diode. It should be a positively clamped signal.
10. Look at the output. It should be a dc voltage with a small ripple. Measure the dc output voltage with a VOM and record in Table 10-2.
11. Switch the oscilloscope to ac input and high sensitivity to measure the ripple. Record V_{rip}.

DC Return

12. In Fig. 10-3, the inside of the dashed box simulates a capacitively coupled source. The 1-kΩ resistor is a dc return. Estimate and record the positive peak output voltage (Table 10-3). Visualize the dc return open; estimate and record the positive-peak output voltage.
13. Connect the circuit of Fig. 10-3. Adjust the source

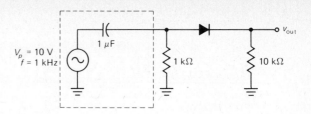

Figure 10-3

to get 1 kHz and 20 V peak-to-peak across the 1-kΩ resistor.
14. Look at the output with the oscilloscope. It should be a half-wave signal. Measure and record the peak value in Table 10-3.
15. Disconnect the dc return. Measure and record the output peak value.

Troubleshooting (Optional)

16. In Fig. 10-2, assume capacitor C_1 is open. Estimate and record the dc output voltage in Table 10-4.
17. Connect the circuit with the foregoing trouble. Measure and record the dc output voltage.
18. Repeat Steps 16 and 17 for each of the remaining troubles listed in Table 10-2.

Design (Optional)

19. The frequency is changed to 2.5 kHz and the load resistor to 47 kΩ in Fig. 10-2. Select a value of output filter capacitance (nearest standard size) that produces a peak-to-peak ripple of approximately 0.1 V. Calculate all quantities listed in Table 10-5.

Computer (Optional)

21. Write and run a program that calculates the dc output voltage and peak-to-peak ripple in Fig. 10-2. The inputs are peak input voltage, frequency, capacitance, and load resistance.

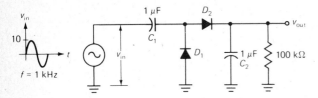

Figure 10-2

DATA FOR EXPERIMENT 10

Table 10-1. Clampers

	Estimated		Measured	
	Pos Peak	Neg Peak	Pos Peak	Neg Peak
Positive clamper				
Negative clamper				

Table 10-2. Peak-to-Peak Detector

	Estimated	Measured
V_{dc}		
V_{rip}		

Table 10-3. DC Return

	Estimated V_P	Measured V_P
With dc return		
Without dc return		

Table 10-4. Troubleshooting

	Estimated V_{dc}	Measured V_{dc}
Open C_1		
Short C_1		
Open C_2		
Short C_2		
Open D_1		
Short D_1		
Open D_2		
Short D_2		

Table 10-5. Design

	Calculated	Measured
Capacitance		
DC output voltage		
Ripple frequency		
Peak-to-peak ripple		

QUESTIONS FOR EXPERIMENT 10

1. If the diode of Fig. 10-1 is reversed, the output will be: ( )
   (a) positively clamped; (b) negative clamped: (c) half-wave rectified;
   (d) peak rectified.
2. If V_P is 10 V in Fig. 10-2, the maximum positive voltage across the first ( )
   diode is approximately:
   (a) 5 V; (b) 10 V; (c) 15 V; (d) 20 V.
3. If V_P is 10 V in Fig. 10-2, the dc output voltage is ideally: ( )
   (a) 0 V; (b) 5 V; (c) 10 V; (d) 20 V.
4. In Fig. 10-2, the peak-to-peak ripple is approximately what percent of dc ( )
   output voltage?
   (a) 0; (b) 1%; (c) 5%; (d) 63.6%.
5. When the dc return of Fig. 10-3 is disconnected, which of the following is ( )
   true?
   (a) The capacitor charges to approximately $2V_P$; (b) current flows easily
   in the reverse diode direction; (c) the diode conducts briefly near each
   positive peak; (d) the diode eventually stops conducting.
6. Explain how a positive dc clamper works.

Troubleshooting (Optional)

7. What dc output voltage did you get when C_2 was opened in Fig. 10-2? Explain why
   this happened.

8. A group of technicians is gathered around a circuit that works with one signal
   generator but not with another. No one can figure out why one generator works but
   not the other. Explain what is probably happening.

Design (Optional)

9. A design rule for a peak-to-peak detector is to make the discharging time constant at least 100 times the period of the input signal. Satisfying this rule produces less than 1 percent peak-to-peak ripple. Derive this rule from Eq. (4-8) in the textbook.

10. Optional. Instructor's question.

EXPERIMENT 11

THE ZENER DIODE

▼

Ideally, a zener diode is equivalent to a dc source when operating in the breakdown region. To a second approximation, it is like a dc source with a small internal impedance. Its main advantage is the approximately constant voltage appearing across it. In this experiment you will get data for the zener voltage and zener resistance.

REQUIRED READING

Chapter 5 (Secs. 5-1 and 5-10) of *Electronic Principles*, 5th ed.

EQUIPMENT

1 power supply: adjustable from approximately 0 to 15 V
1 zener diode: 1N753
1 ½-W resistor: 180 Ω
1 VOM (analog or digital multimeter)

PROCEDURE

Zener Voltage

1. Measure the diode's forward and reverse resistance on one of the middle resistance ranges. The reverse/forward resistance ratio should be at least 1000:1.
2. The 1N753 has a nominal zener voltage of 6.2 V. In Fig. 11-1, estimate and record the output voltage for each input voltage listed in Table 11-1.
3. Connect the circuit of Fig. 11-1. Measure and record the output voltage for each input voltage of Table 11-1.

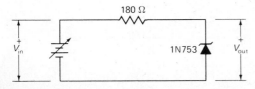

Figure 11-1

Zener Resistance

4. With the data of Table 11-1, calculate and record the zener current in Fig. 11-1 for each entry of Table 11-2.
5. With Eq. (5-12) in the textbook, calculate the zener resistance for $V_{in} = 10$ V. (Use the voltage and current changes between 8 and 12 V.)
6. Calculate and record the zener resistance for $V_{in} = 12$ V.

Curve Tracer

7. If a curve tracer is available, display the forward and reverse zener curves.

Troubleshooting (Optional)

8. In Fig. 11-1, assume V_{in} is 15 V and estimate the output voltage for a shorted zener diode. Record your answer in Table 11-3.
9. Estimate and record the output voltage for an open zener diode.
10. Estimate and record V_{out} for an open resistor.
11. Assume the polarity of the zener diode is reversed. Estimate and record the output voltage for this trouble.
12. Connect the circuit with each of the foregoing troubles. Measure and record V_{out} for a V_{in} of 15 V.

Design (Optional)

13. Select a current-limiting resistor to produce a zener current of approximately 16.5 mA when V_{in} is 14 V.

Record your design value at the top of Table 11-4. Connect the circuit with your design value of R_S. Measure and record the output voltage for each input voltage listed in Table 11-4.

14. Calculate and record the zener current for each input voltage of Table 11-4. Calculate and record the zener resistance for an input voltage of 12 V.

Computer (Optional)

15. Write and run a program that calculates the zener resistance for an input voltage of 10 V in Fig. 11-1. Your program should use some voltages and currents from Tables 11-1 and 11-2.

DATA FOR EXPERIMENT 11

Table 11-1. Data for Zener Diode

V_{in}	Estimated V_{out}	Measured V_{out}
0		
2 V		
4 V		
6 V		
8 V		
10 V		
12 V		
14 V		

Table 11-2. Zener Resistance

V_{in}	Calculated I_Z	Calculated R_Z
0		(no entry)
2 V		(no entry)
4 V		(no entry)
6 V		(no entry)
8 V		(no entry)
10 V		
12 V		
14 V		(no entry)

Table 11-3. Troubleshooting

Trouble	Estimated V_{dc}	Measured V_{dc}
Shorted diode		
Open diode		
Open resistor		
Reversed diode		

Table 11-4. Design: R_S = _____

V_{in}	Measured V_{out}	Calculated I_Z	Calculated R_Z
10 V			(no entry)
12 V			
14 V			(no entry)

QUESTIONS FOR EXPERIMENT 11

1. In Fig. 11-1, the zener current and the current through the 180-Ω resistor   ( )
   are:
   (a) equal; (b) almost equal; (c) much different.
2. The zener diode starts to break down when the input voltage is approximately: ( )
   (a) 4 V; (b) 6 V; (c) 8 V; (d) 10 V.
3. When V_{in} is less than 6 V, the output voltage is: ( )
   (a) approximately constant; (b) 6 negative; (c) the same as the input.
4. When V_{in} is greater than 8 V, the output voltage is: ( )
   (a) approximately constant; (b) negative; (c) the same as the input.
5. The calculated zener resistances were closest to: ( )
   (a) 1 Ω; (b) 2 Ω; (c) 7 Ω; (d) 20 Ω.
6. Explain why the zener diode is called a constant-voltage device.

Troubleshooting (Optional)

7. Explain the value of output voltage you got when the zener diode was open.

8. Explain the value of output voltage you got when the zener diode was reversed.

Design (Optional)

9. How and why did you select the value of current-limiting resistance?

10. Optional. Instructor's question.

EXPERIMENT 12

THE ZENER REGULATOR

▼

In a zener voltage regulator a load resistor is in parallel with a zener diode. As long as the zener diode operates in the breakdown region, the load voltage is approximately constant and equal to the zener voltage. In a stiff zener regulator, the zener resistance is less than 1/100 of the series resistance and less than 1/100 of the load resistance. By meeting the first condition, a zener regulator attenuates the input ripple by a factor of at least 100. By meeting the second condition, a zener regulator appears like a stiff voltage source to the load resistance.

In this experiment you will build a split-supply with regulated positive and negative output voltages. This will allow you to verify the operation of a zener regulator as described in your textbook.

REQUIRED READING

Chapter 5 (Sec. 5-2) of *Electronic Principles*, 5th ed.

EQUIPMENT

1 center-tapped transformer, 12.6 V ac (Triad F-25X or equivalent) with fused line cord
4 silicon diodes: 1N4001 (or equivalent)
3 zener diodes: 1N753
8 ½-W resistors: two 150 Ω, two 470 Ω, two 4.7 kΩ, two 47 kΩ
2 capacitors: 470 μF (25-V rating or better)

1 VOM (analog or digital multimeter)
1 oscilloscope

PROCEDURE

Split Supply

1. A 1N753 has a nominal zener voltage of 6.2 V. In Fig. 12-1, calculate the input and output voltages for each zener regulator. (The input voltages are across the filter capacitors.) Record your answers in Table 12-1.

2. Connect the split-supply of Fig. 12-1.

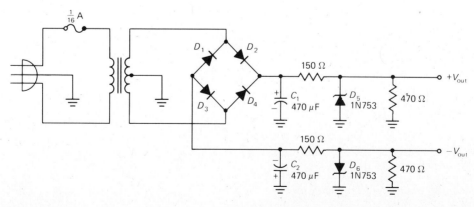

Figure 12-1

3. Measure the input and output voltages of each zener regulator. Record your data in Table 12-1.

Voltage Regulation

4. Estimate and record the output voltages in Fig. 12-1 for each of the load resistors listed in Table 12-2.
5. Connect the circuit. Measure and record the output voltages for the load resistances of Table 12-2.

Ripple Attenuation

6. For each load resistance listed in Table 12-3, calculate and record the peak-to-peak ripple across the upper filter capacitor of Fig. 12-1. Also calculate and record the peak-to-peak ripple at the positive output. (Assume a zener resistance of 7 Ω.)
7. For each load resistance of Table 12-3, measure and record the peak-to-peak ripple at the input and output of the positive zener regulator.

Troubleshooting (Optional)

8. Assume the center tap of Fig. 12-1 is open.
9. Estimate the output voltages for the foregoing trouble. Record your answers in Table 12-4.
10. Connect the circuit with the foregoing trouble. Measure and record the output voltages. Remove the trouble.
11. Repeat Steps 9 and 10 for the other troubles listed in Table 12-4.

Design (Optional)

12. Design a two-stage voltage regulator similar to Fig. 5-8 of your textbook to meet these specifications: preregulator output is a nominal +12.4 V, final output is a nominal +6.2 V, current in preregulator series resistor is 40 mA, current in final series resistor is 20 mA, and ripple attenuation is at least 300. Assume a zener resistance of 7 Ω for each diode. Use three 1N753s and any additional resistors as required. Draw your final design at the bottom of Table 12-5.

13. Calculate and record the dc voltage and peak-to-peak ripple at the preregulator input, regulator input, and final output (Table 12-5).

14. Check with the instructor about the safety of your design. Then connect your design with a load resistance of 470 Ω. Measure all dc voltages and ripples listed in Table 12-5. Record your data.

Computer (Optional)

15. Enter and run this program:

```
10 PRINT "ENTER ZENER VOLTAGE"
20 INPUT VZ
30 PRINT "ENTER LOAD RESISTANCE"
40 INPUT RL
50 IL = VZ/RL
60 PRINT "THE LOAD CURRENT IS"
70 PRINT IL
80 GOTO 10
```

16. Write and run a program that prints out the value of zener current after you input the values of V_S, V_Z, R_S, and R_L. Include a GOTO statement to repeat the program.

DATA FOR EXPERIMENT 12

Table 12-1. Split Supply

	Calculated		Measured	
	V_{in}	V_{out}	V_{in}	V_{out}
Positive regulator				
Negative regulator				

Table 12-2. Voltage Regulation

	Estimated		Measured	
R_L	$+V_{out}$	$-V_{out}$	$+V_{out}$	$-V_{out}$
470 Ω				
4.7 kΩ				
47 kΩ				

Table 12-3. Ripple

	Calculated V_{rip}		Measured V_{rip}	
R_L	In	Out	In	Out
470 Ω				
47 kΩ				

Table 12-4. Troubleshooting

	Estimated		Measured	
	$+V_{out}$	$-V_{out}$	$+V_{out}$	$-V_{out}$
Open CT				
Open D_1				
Open D_6				

Table 12-5. Design

	Calculated		Measured	
	V_{dc}	V_{rip}	V_{dc}	V_{rip}
Preregulator input				
Regulator input				
Regulator output				

Draw your design here:

QUESTIONS FOR EXPERIMENT 12

1. A split supply has: (   )
   (a) only one output voltage;   (b) only a positive output voltage;
   (c) only a negative output voltage;   (d) positive and negative outputs.
2. The value of V_{in} to the positive zener regulator is closest to: (   )
   (a) 5 V;   (b) 10 V;   (c) 15 V;   (d) 20 V.
3. When R_L increases in Table 12-2, the measured positive output voltage: (   )
   (a) decreases slightly: (b) remains the same;   (c) increases slightly.
4. Theoretically, the positive zener regulator of Fig. 12-1 attenuates the ripple   (   )
   by a factor of approximately:
   (a) 10;   (b) 20;   (c) 50;   (d) 100.
5. The current through either series resistor of Fig. 12-1 is closest to: (   )
   (a) 5 mA;   (b) 10 mA;   (c) 15 mA;   (d) 20 mA.
6. Explain how the positive zener regulator of Fig. 12-1 works.

Troubleshooting (Optional)

7. Explain why the circuit of Fig. 12-1 continued to work even though you opened the
   center tap.

8. Explain why the circuit of Fig. 12-1 still works with an open D_2.

Design (Optional)

9. Explain why the measured ripples did not agree exactly with the calculated ripples in
   your design.

10. Optional. Instructor's question.

EXPERIMENT 13

OPTOELECTRONIC DEVICES

▼

In a forward-biased LED, heat and light are radiated when free electrons and holes recombine at the junction. Because the LED material is semi-transparent, some of the light escapes to the surroundings. LEDs have a typical voltage drop from 1.5 and 2.5 V for currents between 10 and 50 mA. The exact voltage drop depends on the color, tolerance, and other factors. For troubleshooting and design, we will use the second diode approximation with a knee of 2 V.

An LED array is a group of LEDs that display numbers, letters, or other symbols. The most common LED array is the seven-segment display. It contains seven rectangular LEDs. Each LED is called a segment because it forms one part of the character being displayed. By activating one or more LEDs, we can form any digit from 0 through 9.

An optocoupler combines an LED and a photodetector in a single package. The light from the LED hits the photodetector. This produces an output voltage that depends on the amount of current through the LED. If the LED current has an ac variation, V_{out} will have an ac variation. The key advantage of an optocoupler is the electrical isolation between the LED circuit and the output circuit, typically in thousands of megohms.

REQUIRED READING

Chapter 5 (Sec. 5-3) of *Electronic Principles*, 5th ed.

EQUIPMENT

2 power supplies: one at 15 V, another adjustable from approximately 0 to 15 V
2 LEDs: TIL221 and TIL222 (or equivalent red and green LEDs)
2 1-W resistors: 270 Ω
Seven-segment display: TIL312 (or equivalent)
Optocoupler: 4N26 (or equivalent)
1 VOM (analog or digital multimeter)

PROCEDURE

Data for a Red LED

1. Examine the red LED. Notice that one side of the package has a flat edge. This indicates the cathode side. (With many LEDs, the cathode lead is slightly shorter than the anode lead. This shorter lead is another way to identify the cathode.)
2. Connect the circuit of Fig. 13-1 using a red LED.
3. Adjust source voltage V_S to get 10 mA through the LED. Record the corresponding LED voltage in Table 13-1.
4. Adjust the source voltage and set up the remaining currents listed in Table 13-1. Record each LED voltage.

Data for a Green LED

5. Replace the red LED by a green LED in the circuit of Fig. 13-1.
6. Repeat Steps 3 and 4 for the green LED.

Using a Seven-Segment Display

7. Figure 13-2a shows the pinout for the seven-segment display used in this experiment (top view). It

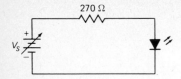

Figure 13-1

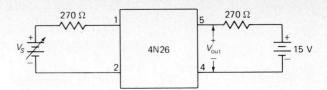

Figure 13-3

includes a left decimal point (LDP) and a right decimal point (RDP). Connect the circuit of Fig. 13-2*b*.

8. Figure 13-2*c* shows the schematic diagram for a TIL312. Ground pins 1, 10, and 13. If the circuit is working correctly, digit 7 will be displayed.

9. Disconnect the grounds on pins 1, 10, and 13.

10. Refer to Fig. 13-2*a* and *c*. Which pins should you ground to display a zero? Ground these pins and if the circuit is working correctly, enter the pin numbers in Table 13-2.

11. Repeat Step 10 for the remaining digits, 1 through 9, and the decimal points.

The Transfer Graph of an Optocoupler

12. Connect the circuit of Fig. 13-3. Adjust the source voltage to 2 V. Measure and record the output voltage (Table 13-3).

13. Repeat Step 12 for the source voltages shown in Table 13-3. Record the corresponding output voltages.

Troubleshooting (Optional)

14. If V_S is 15 V in Fig. 13-1, estimate the voltage across the red LED if it is open. Record your answer in

Table 13-4. Similarly, estimate the LED voltage if the LED is shorted.

15. Connect the circuit with a source voltage of 15 V and a red LED. Measure and record the LED voltage for the troubles listed in Table 13-4.

Design (Optional)

16. Select a current-limiting resistor for a red LED in Fig. 13-1 that sets up a current of approximately 20 mA when the source voltage is 15 V. Record your design value in Table 13-5. Calculate and record the LED current and voltage.

17. Connect your design. Measure and record the LED current and voltage.

18. Repeat Steps 16 and 17 for the green LED.

Computer (Optional)

19. Write and run a program that calculates the LED current in Fig. 13-1 for a source voltage of 15 V and an LED voltage drop of 2 V.

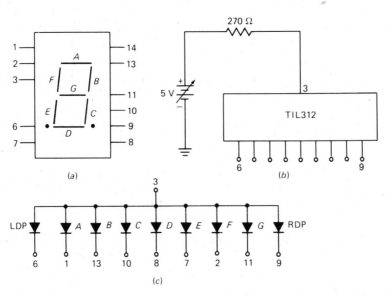

Figure 13-2

DATA FOR EXPERIMENT 13

Table 13-1. LED Data

I	V_{red}	V_{green}
10 mA		
20 mA		
30 mA		
40 mA		

Table 13-2. Seven-Segment Indicator

Display	Pins grounded
0	
1	
2	
3	
4	
5	
6	
7	
8	
9	
LDP	
RDP	

Table 13-3. Optocopler

V_S	V_{out}
2 V	
4 V	
6 V	
8 V	
10 V	
12 V	
14 V	

Table 13-4. Troubleshooting

	Estimated V_{LED}	Measured V_{LED}
Open LED		
Shorted LED		

Table 13.5. Design

		Calculated		Measured	
	R	I_{LED}	V_{LED}	I_{LED}	V_{LED}
Red LED					
Green LED					

QUESTIONS FOR EXPERIMENT 13

1. The voltage drop across the red LED for a current of 30 mA was closest to: (   )
   (**a**) 0 V;   (**b**) 1 V;   (**c**) 2 V;   (**d**) 4 V.
2. The voltage drop across the green LED for a current of 30 mA was closest (   )
   to:
   (**a**) 0 V;   (**b**) 1 V;   (**c**) 2 V;   (**d**) 4 V.
3. To display a 1 on the seven-segment indicator, which pins did you ground? (   )
   (**a**) 1;   (**b**) 1 and 10;   (**c**) 10 and 13;   (**d**) 2, 7, and 8.
4. In Fig. 13-2, which of the following is true? (   )
   (**a**) LED brightness decreases as more segments are lit;   (**b**) all segments
   are equally bright at all times;   (**c**) number 8 was brighter than number 1.
5. When the source voltage increases in Fig. 13-3, the output voltage: (   )
   (**a**) decreases;   (**b**) stays the same;   (**c**) increases.
6. Explain how the LED array of Fig. 13-2 works. Include a discussion of the LED
   current versus the total current.

Troubleshooting (Optional)

7. Why was the LED voltage large when the LED was open?

Design (Optional)

8. Explain how you calculated the current-limiting resistor for the red LED in Table 13-5.

9. It is possible to get equal brightness of all numbers with the LED array of Fig. 13-2. How can this be done?

10. Optional. Instructor's question.

EXPERIMENT 14

THE CE CONNECTION

▼

As an approximation of transistor behavior, we use the Ebers-Moll model: the emitter diode acts like a rectifier diode, while the corrector diode acts like a controlled-current source. The voltage across the emitter diode of a small-signal transistor is typically 0.6 to 0.7 V. For most troubleshooting and design, we will use 0.7 V for the V_{BE} drop. In this experiment, you will get data for calculating the α_{dc}, β_{dc}, and the V_{BE} drop.

When the maximum ratings of a transistor are exceeded, it can be damaged in several ways. The most common transistor trouble is a collector-emitter short where both the emitter diode and the collector diode are shorted. Another common transistor trouble is the collector-emitter open where both the emitter diode and the collector diode are open. Besides the foregoing, it is possible to have only one diode shorted, only one diode open, a leaky diode, etc.

To keep the troubleshooting straightforward, we will emphasize the two most common troubles: the collector-emitter short and the collector-emitter open. We will simulate a collector-emitter short by putting a jumper between the collector, base, and emitter; this shorts all three terminals together. We will simulate the collector-emitter open by removing the transistor from the circuit; this opens both diodes.

REQUIRED READING

Chapter 6 (Secs. 6-1 to 6-9, and 6-11) of *Electronic Principles*, 5th ed.

EQUIPMENT

1 power supply: 15 V
3 ½-W resistors: 100 Ω, 1 kΩ, 470 kΩ
3 transistors: 2N3904 (or almost any small-signal *npn* silicon transistor)
1 VOM (analog or digital multimeter)

PROCEDURE

Ohmmeter Tests

1. Measure the resistance between the collector and emitter of one of the transistors. This resistance should be extremely high (hundreds of megohms) in either direction.
2. Measure the forward and reverse resistance of the base-emitter diode and the collector-base diode. For both diodes, the reverse/forward resistance ratio should be at least 1000:1.
3. Repeat Steps 1 and 2 for the two other transistors.

Transistor Characteristics

4. Connect the circuit of Fig. 14-1, using one of the transistors.
5. Measure and record V_{BE} and V_{CE} in Table 14-1.
6. Measure and record I_C and I_B in Table 14-1.
7. Calculate the values of V_{CB}, I_E, α_{dc}, and β_{dc} in Fig. 14-1. Record in Table 14-2.
8. Repeat Steps 4 to 7, using a second transistor.
9. Repeat Steps 4 to 7, using a third transistor.
10. If a curve tracer is available, display the collector

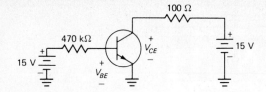

Figure 14-1

curves of all three transistors. Notice the differences in β_{dc}, breakdown voltages, etc.

Troubleshooting (Optional)

11. In Fig. 14-2, estimate and record the collector-to-ground voltage V_C for each trouble listed in Table 14-3. Note: To simulate a collector-emitter short, put a jumper between the collector, emitter, and base so that all three terminals are shorted together. To simulate a collector-emitter open, remove the transistor from the circuit.

12. Connect the circuit with each of the foregoing troubles. Measure and record the collector voltage for each trouble.

Design (Optional)

13. Connect the circuit of Fig. 14-2 and measure the collector-to-ground voltage.

14. With the data of Step 13, calculate β_{dc} and select a base resistance that will produce a collector voltage of approximately half the supply voltage. Record the

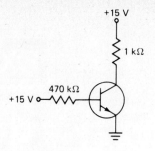

Figure 14-2

nearest standard resistance and the calculated collector voltage in Table 14-4.

15. Connect the circuit with your design value of base resistance. Complete the entries of Table 14-4.

Computer (Optional)

16. Enter and run this program for Fig. 14-2:

```
10 PRINT "ENTER COLLECTOR SUPPLY
   VOLTAGE": INPUT VCC
20 PRINT "ENTER COLLECTOR RESIS-
   TANCE": INPUT RC
30 IC = VCC/RC
40 PRINT "THE COLLECTOR SATURATION
   CURRENT IS": PRINT IC
50 END
```

17. Write and run a program that prints out the collector current for Fig. 14-2.

DATA FOR EXPERIMENT 14

Table 14-1. Transistor Voltages and Currents

Transistor	V_{BE}	V_{CE}	I_B	I_C
1				
2				
3				

Table 14-2. Calculations

Transistor	V_{CB}	I_E	α_{dc}	β_{dc}
1				
2				
3				

Table 14-3. Troubleshooting

Trouble	Estimated V_C	Measured V_C
Open 470 kΩ		
Shorted 1 kΩ		
Open 1 kΩ		
Shorted collector-emitter		
Open collector-emitter		

Table 14-4. Design

	R_B	V_C
Calculated		
Measured		

QUESTIONS FOR EXPERIMENT 14

1. The V_{BE} drop of the transistors was closest to: ( )
   (**a**) 0 V; (**b**) 0.3 V; (**c**) 0.7 V; (**d**) 1 V.
2. The α_{dc} of all transistors was very close to: ( )
   (**a**) 0; (**b**) 1; (**c**) 5; (**d**) 20.
3. The β_{dc} of all transistors was greater than: ( )
   (**a**) 0; (**b**) 1; (**c**) 5; (**d**) 20.
4. This experiment proves that collector current is much greater than: ( )
   (**a**) collector voltage; (**b**) emitter current; (**c**) base current; (**d**) 0.7 V.

5. The transistors were silicon because: ( )
   (a) V_{BE} was approximately 0.7 V;   (b) I_C is much greater than I_B;
   (c) the collector diode was reverse-biased;   (d) β_{dc} was much greater than unity.

6. What did you learn about the V_{BE} drop and the relation of collector current to base current?

Troubleshooting (Optional)

7. What collector voltage did you measure when the base resistor was open? Explain why this voltage existed at the collector.

8. Briefly explain why the collector voltage is approximately zero when a transistor has a collector-emitter short.

Design (Optional)

9. Explain how you calculated the base resistance in Table 14-4.

10. Optional. Instructor's question.

EXPERIMENT 15

COLLECTOR CURVES

▼

Each base current in a transistor produces a different collector curve. The collector curve is a graph of collector current versus collector-emitter voltage. In this experiment, you will build a circuit that simulates the action of a curve tracer. The idea is to use the vertical and horizontal inputs of an oscilloscope to monitor the collector current and the collector-emitter voltage. By varying the collector supply voltage, you will be able to create collector curves on the oscilloscope.

REQUIRED READING

Chapter 6 (Sec. 6-6) of *Electronic Principles*, 5th ed.

EQUIPMENT

2 power supplies: 9 V and adjustable from 1 to 10 V
1 transistor: 2N3904 (or almost any small-signal *npn* silicon transistor)
2 ½-W resistors: 100 Ω, 100 kΩ
1 decade resistance box (or substitute 220 kΩ, 470 kΩ, 1 mΩ)
1 oscilloscope

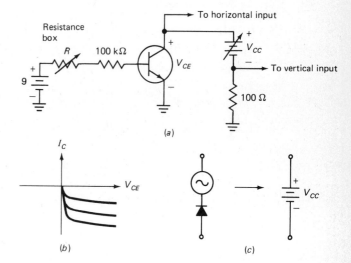

Figure 15-1

PROCEDURE

1. Set up the oscilloscope by turning the horizontal sensitivity control to 1 V/cm (dc input) and the vertical sensitivity control to 0.1 V/cm (dc input). Center the undeflected spot in the upper left-hand corner of the screen.
2. Connect the circuit of Fig. 15-1a. Notice the collector current passes through the 100-Ω resistor. Because of this, the voltage to the vertical input is proportional to I_C. In fact, each milliampere of collector current produces 100 mV of vertical input.
3. Set the resistance box R to 1 MΩ.
4. Vary the V_{CC} supply back and forth rapidly between minimum and 10 V. You will see the first collector curve of Fig. 15-1b.

5. Change R to 470 kΩ. Again vary the V_{CC} rapidly between low and high voltage. You should see the second curve of Fig. 15-1b.
6. Change R to 220 kΩ and vary V_{CC} rapidly. You should see a third collector curve.
7. Return R to 1 MΩ. Set the V_{CC} supply to produce each V_{CE} listed in Table 15-1. For each V_{CE}, read and record the value of I_C.
8. In a similar way, fill in the rest of Table 15-1 for the other values of R.
9. This step and the next two are optional; ask your

instructor. If an audio generator with a *floating output* is available, connect a diode in series with it as shown in Fig. 15-1c. Then substitute this generator-diode combination for the I_{CC} supply of Fig. 15-1a.

10. Adjust the audio generator to 100 Hz and enough signal to produce a collector curve. Each time you change the resistor substitution box, you get a different collector curve.

11. If a transistor curve tracer is available, use it to look at the collector curves of your transistor.

DATA FOR EXPERIMENT 15

Table 15-1. Collector Curves

		0	2	4	6	8	10
				V_{CE}			
$R = 1$ MΩ	I_C						
$R = 470$ kΩ	I_C						
$R = 220$ kΩ	I_C						

QUESTIONS FOR EXPERIMENT 15

1. When R in Fig. 15-1a equals 1 MΩ, base current is closest to: ( )
   (a) 2 μA;   (b) 4 μA;   (c) 8 μA;   (d) 16 μA.
2. If the voltage across the 100 Ω resistor of Fig. 15-1a is 100 mV when R ( )
   equals 1 MΩ, β_{dc} is closest to:
   (a) 40;   (b) 80;   (c) 120;   (d) 180.
3. Collector curves appear upside-down in this experiment because: ( )
   (a) the transistor is *npn*;   (b) the voltage across the 100-Ω resistor is
   negative;   (c) negative end of $\mathbf{V}_{CC}$ is connected to vertical input;   (d) the
   voltage across the 100-Ω resistor is positive.
4. The I_C values of Table 15-1 prove: ( )
   (a) collector current is constant;   (b) the transistor breaks down;   (c) base
   current is constant;   (d) I_C is almost constant when V_{CE} is between 2 and
   10 V.
5. Because of the values in Table 15-1, $V_{CE(sat)}$ must be: ( )
   (a) less than zero;   (b) less than 2 V;   (c) more than 2 V;   (d) more than
   10 V.
6. Table 15-1 proves collector breakdown voltage is: ( )
   (a) less than zero;   (b) less than 2 V;   (c) more than 2 V;   (d) more than
   10 V.
7. In Fig. 15-1a, a collector current of 2 mA produces a vertical deflection of: ( )
   (a) 0;   (b) 1 cm;   (c) 2 cm;   (d) 5 cm.
8. In Fig. 15-1a, a collector-emitter voltage of 5 V produces a horizontal deflec- ( )
   tion of:
   (a) 0;   (b) 1 cm;   (c) 2 cm;   (d) 5 cm.

9. Optional. Instructor's question.

10. Optional. Instructor's question.

70

EXPERIMENT 16

BASE BIAS

▼

A circuit like Fig. 16-1 is referred to as *base bias*, because it sets up a fixed base current. You can calculate the base current by applying Ohm's law to the total base resistance. This base current will remain constant when you replace transistors.

On the other hand, the collector current equals the current gain times the base current. Because of this, the collector current may have large variations from one transistor to the next. In other words, the Q point in a base-biased circuit is heavily dependent on the value of β_{dc}.

REQUIRED READING

Chapter 7 (Secs. 7-1 to 7-5) of *Electronic Principles*, 5th ed.

EQUIPMENT

1 power supply: 15 V
3 transistors: 2N3904 (or almost any small-signal *npn* silicon transistor)
2 ½-W resistors: 2.2 kΩ and 22 kΩ
1 decade resistance box (or substitute a 1-MΩ potentiometer)

PROCEDURE

1. The fixed-base-current circuit of Fig. 16-1 is not a stable biasing circuit, but it is a good way to measure β_{dc}.
2. Connect the circuit of Fig. 16-1 using one of the transistors.
3. Adjust R to get a V_{CE} of 1 V. Record the value of R in Table 16-1. (If R is a potentiometer instead of a decade box, you will have to disconnect it and measure its resistance.) In Fig. 16-1, notice the total base

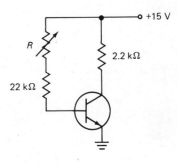

Figure 16-1

resistance R_B equals R plus 22 kΩ. Record the value of R_B in Table 16-1.
4. Calculate the values of β_{dc} and I_C. Record in Table 16-1.
5. Repeat Steps 2 through 4 for the second and third transistors.
6. With the values of Table 16-1, calculate the ideal and second-approximation values of I_E in Fig. 16-1. Record the I_E values in Table 16-2.
7. If a curve tracer or other transistor tester is available, measure the β_{dc} of each transistor for an I_C of approximately 5 mA. The values should be similar to the β_{dc} values of Table 16-1.

Name _____ Date _____

DATA FOR EXPERIMENT 16

Table 16-1. β_{dc} **Values**

Transistor	R	R_B	β_{dc}	I_C
1				
2				
3				

Table 16-2. Calculations

Test	$I_{E(\text{ideal})}$	$I_{E(\text{second})}$
1		
2		
3		

QUESTIONS FOR EXPERIMENT 16

1. In Fig. 16-1, an increase in current gain causes an increase in: ( )
   (a) I_C; (b) V_{CE}; (c) I_B; (d) V_{CC}.
2. When R_B increases in a base-biased circuit, which of these increases? ( )
   (a) I_E; (b) V_{BB}; (c) I_C; (d) V_{CE}.
3. When R_C increases in a base-biased circuit, which of these decreases? ( )
   (a) I_E; (b) V_{BB}; (c) I_C; (d) V_{CE}.
4. When V_{BB} increases in a base-biased circuit, which of these decreases? ( )
   (a) I_B; (b) V_{BE}; (c) I_C; (d) V_{CE}.
5. When β_{dc} increases in a base-biased circuit, which of these remains constant? ( )
   (a) I_B; (b) I_E; (c) I_C; (d) V_{CE}.
6. The ideal and second-approximation values in Table 16-2 differ by approxi- ( )
   mately:
   (a) 0.1%; (b) 1%; (c) 5%; (d) 10%.
7. If the base current is 10 μA in Fig. 16-1 and the collector voltage is 10 V, ( )
   the current gain is closest to:
   (a) 50; (b) 125; (c) 225; (d) 350.
8. If the collector voltage is 5 V in Fig. 16-1 and β_{dc} is 150, the base current is ( )
   closest to:
   (a) 10 μA; (b) 20 μA; (c) 30 μA; (d) μA.

9. Optional. Instructor's question.

10. Optional. Instructor's question.

EXPERIMENT 17

LED DRIVERS

▼

The simplest way to use a transistor is as a switch, meaning that it operates at either saturation or cutoff but nowhere else along the load line. When saturated, a transistor appears as a closed switch between its collector and emitter terminals. When cut off, it is like an open switch. Because of the wide variation in β_{dc}, hard saturation is used with transistor switches. This means having enough base current to guarantee transistor saturation under all operating conditions. With small-signal transistors, hard saturation requires a base current of approximately one-tenth of the collector saturation current.

Another basic way to use the transistor is as a direct current source. In this case, the base resistor is omitted and the base supply voltage is connected directly to the base terminal. To set up the desired collector current, we use an emitter resistor. The emitter is bootstrapped to within one V_{BE} drop of the base voltage. Therefore, the collector current equals $(V_{BB} - V_{BE})$ divided by R_E. This fixed collector current then flows through the load, which is connected between the collector and positive supply voltage.

In this experiment you will build a transistor switch and a transistor current source. You will also have the opportunity to troubleshoot and design these basic transistor circuits.

REQUIRED READING

Chapter 7 (Secs. 7-1 to 7-9) of *Electronic Principles*, 5th ed.

EQUIPMENT

2 power supplies: each source adjustable from 0 to 15 V
3 ½-W resistors: 220 Ω, 1 kΩ, 10 kΩ
1 LED: TIL221 (or an equivalent red LED)
3 transistors: 2N3904
1 VOM (analog or digital multimeter)

PROCEDURE

1. In Fig. 17-1, calculate I_B, I_C, and V_{CE}. Record your answers in Table 17-1.

2. Connect the transistor switch of Fig. 17-1. Measure and record the quantities listed in Table 17-1.
3. Repeat Steps 1 and 2 for the other transistors.

Transistor Current Source

4. In Fig. 17-2, calculate all quantities listed in Table 17-2.
5. Connect the transistor current source of Fig. 17-2. Measure and record the quantities listed in Table 17-2.
6. Repeat Steps 4 and 5 for the other transistors.

Troubleshooting (Optional)

7. In Fig. 17-1, assume the base resistor is open. Estimate and record the collector voltage in Table 17-3.

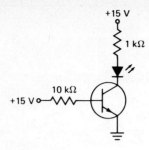

Figure 17-1

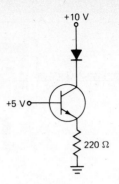

Figure 17-2

8. Repeat Step 7 for each of the troubles listed in Table 17-3.
9. Connect the circuit of Fig. 17-1 with each of the troubles listed in Table 17-3. Measure and record all listed quantities.
10. In Fig. 17-2, assume the emitter resistor is open. Estimate and record the voltages listed in Table 17-4.
11. Repeat Step 10 for each of the troubles listed in Table 17-4.
12. Connect the circuit of Fig. 17-2 with each of the troubles listed in Table 17-4. Measure and record all listed quantities.

Design (Optional)

13. Select a collector resistance in Fig. 17-1 to produce a collector current of approximately 30 mA. Calculate and record the quantities listed in Table 17-5.

14. Connect the circuit of Fig. 17-1 with your design value of collector resistance. Measure and record the quantities of Table 17-5.
15. Select an emitter resistance in Fig. 17-2 to get a collector current of approximately 30 mA. Calculate and record the quantities listed in Table 17-5.
16. Connect the circuit of Fig. 17-2 with your design value of emitter resistance. Measure and record all quantities in Table 17-5.

Computer (Optional)

17. Write and run a program that prints out the collector current for Fig. 17-2.

DATA FOR EXPERIMENT 17

Table 17-1. Transistor Switch

Transistor	I_B	Calculated I_C	V_{CE}	I_B	Measured I_C	V_{CE}
1						
2						
3						

Table 17-2. Transistor Current Source

Transistor	V_E	Calculated I_C	V_{CE}	V_E	Measured I_C	V_{CE}
1						
2						
3						

Table 17-3. Troubleshooting the Transistor Switch

Trouble	Estimated V_C	Measured V_C
Open 10 kΩ		
Open 1 kΩ		
Shorted collector-emitter		
Open collector-emitter		

Table 17-4. Troubleshooting the Transistor Current Source

Trouble	Estimated V_C	V_E	Measured V_C	V_E
Open 220 Ω				
Shorted collector-emitter				
Open collector-emitter				

Table 17-5. Design

Transistor	R	Calculated V_E	I_C	Measured V_E	I_C
Switch					
Current source					

QUESTIONS FOR EXPERIMENT 17

1. In Fig. 17-1, the ratio of collector current to base current is closest to: ( )
   (a) 1; (b) 10; (c) 100; (d) 300.

2. The measured V_{CE} entries of Table 17-1 indicate that collector voltage is ( )
   approximately;
   (a) 0; (b) 2 V; (c) 4 V; (d) 8 V.

3. In the transistor current source of Fig. 17-2, the emitter voltage is closest to: ( )
   (a) 0.7 V; (b) 4.3 V; (c) 5 V; (d) 10 V.

4. When a transistor is in hard saturation, its collector-emitter terminals appear ( )
   approximately:
   (a) shorted; (b) open; (c) in the active region; (d) cut off.

5. With a transistor current source; the emitter is bootstrapped to within one ( )
   V_{BE} drop of the:
   (a) base voltage; (b) emitter voltage; (c) collector voltage; (d) collector current.

6. What are some of the differences between a transistor switch and a transistor current
   source?

Troubleshooting (Optional)

7. While troubleshooting a transistor switch like Fig. 17-1, you notice that the collector
   voltage is always zero. If the LED is lit, what is the most likely trouble?

8. Explain the measured values for collector and emitter voltage when the emitter resistor
   was open in Fig. 17-2.

Design (Optional)

9. Why is hard saturation used with a transistor switch?

10. Optional. Instructor's question.

SETTING UP A STABLE Q POINT

▼

If you want a stable Q point, you will have to use either voltage-divider bias or two-supply emitter bias. With either of these stable biasing methods, the effects of h_{FE} variations are virtually eliminated. Voltage-divider bias requires only a single power supply. This type of bias is also called universal bias, an indication of its popularity. When two supplies are available, two-supply emitter bias can provide as stable a Q point as voltage-divider bias.

In this experiment you will connect both types of bias and verify the stable Q points discussed in your textbook.

REQUIRED READING

Chapter 8 (Secs. 8-1 to 8-7) of *Electronic Principles*, 5th ed.

EQUIPMENT

2 power supplies: 15 V
3 transistors: 2N3904 (or equivalent)
5 ½-W resistors: 1 kΩ, 2.2 kΩ, 3.6 kΩ, 8.2 kΩ, 10 kΩ
1 VOM (analog or digital multimeter)

PROCEDURE

Voltage-Divider Bias

1. In Fig. 18-1, calculate V_B, V_E, and V_C. Record your answers in Table 18-1.
2. Connect the circuit of Fig. 18-1. Measure and record the quantities listed in Table 18-1.
3. Repeat Steps 1 and 2 for the other transistors.

Emitter Bias

4. In Fig. 18-2, calculate V_B, V_E, and V_C. Record your answers in Table 18-2.
5. Connect the emitter-biased circuit of Fig. 18-2. Measure and record the quantities of Table 18-2.
6. Repeat Steps 4 and 5 for the other transistors.

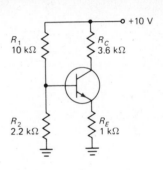

Figure 18-1

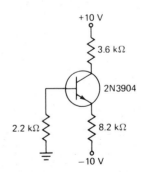

Figure 18-2

Troubleshooting (Optional)

7. In Fig. 18-1, assume the R_1 is open. Estimate and record the collector voltage V_C in Table 18-3.

8. Repeat Step 7 for the other troubles listed in Table 18-3. Connect the circuit of Fig. 18-1 with each trouble listed in Table 18-3. Measure and record the collector voltage.

Design (Optional)

9. Design a stiff voltage-divider biased circuit to meet the following specifications: $V_{CC} = 15$ V, $I_C = 2$ mA, and $V_C = 7.5$ V. You may assume an h_{FE} of 200. Calculate and record the quantities listed in Table 18-4.

10. Connect your design. Measure and record the quantities of Table 18-4.

Computer (Optional)

11. Write and run a program that calculates the I_C and V_{CE} for a voltage-divider bias and emitter bias. Use a menu and include INPUT statements to enter the required data.

DATA FOR EXPERIMENT 18

Table 18-1. Voltage-Divider Bias

Transistor	V_B	Calculated V_E	V_C	V_B	Measured V_E	V_C
1						
2						
3						

Table 18-2. Emitter Bias

Transistor	V_B	Calculated V_E	V_C	V_B	Measured V_E	V_C
1						
2						
3						

Table 18-3. Troubleshooting

Trouble	Estimated V_C	Measured V_C
Open R_1		
Shorted R_1		
Open R_2		
Shorted R_2		
Open R_C		
Shorted R_C		
Open R_E		
Shorted R_E		
Open collector-emitter		
Shorted collector-emitter		

Table 18-4. Design

Values: R_1 = _____ ; R_2 = _____ ; R_C = _____ ; R_E = _____

Transistor	Calculated V_C	Measured V_C
1		
2		
3		

QUESTIONS FOR EXPERIMENT 18

1. Ideally, the voltage divider of Fig. 18-1 produces which of the following base ( )
   voltages:
   (a) 0 V; (b) 1.1 V; (c) 1.8 V; (d) 6.03 V.
2. The measured emitter voltage of Fig. 18-1 was closest to: ( )
   (a) 0 V; (b) 1.1 V; (c) 1.8 V; (d) 6.03 V.
3. The measured collector voltage of Fig. 18-1 was closest to: ( )
   (a) 0 V; (b) 1.1 V; (c) 1.8 V; (d) 6.03 V.
4. The base voltage measured in Fig. 18-2 was: ( )
   (a) 0 V; (b) slightly positive; (c) slightly negative; (d) -0.7 V.
5. With both voltage-divider bias and emitter bias, the measured collector volt- ( )
   age was approximately:
   (a) constant; (b) negative; (c) unstable; (d) one V_{BE} drop less than
   the base voltage.
6. What did you learn about the Q point of a circuit that uses voltage-divider bias or
   emitter bias?

Troubleshooting (Optional)

7. Name all the troubles you found that produced a collector voltage of 10 V.

8. What collector voltage did you measure with a shorted collector-emitter? Explain why
   this value occurred.

Design (Optional)

9. Compare the measured V_C with the calculated V_C in Table 18-4. Explain why the
   measured and calculated values differ.

10. Optional. Instructor's question.

EXPERIMENT 19

BIASING *PNP* TRANSISTORS

▼

Since the emitter and collector diodes of a *pnp* transistor point in the opposite direction of an *npn* transistor, all currents and voltages are reversed in the *pnp* transistor. If only positive power supplies are available, you have to connect the *pnp* transistor upside down. In this experiment you will build *pnp* biasing circuits that work with positive or negative supply voltages.

REQUIRED READING

Chapter 8 (Sec. 8-5) of *Electronic Principles*, 5th ed.

EQUIPMENT

1 power supply: 15 V
3 *pnp* transistors: 2N3906 (or equivalent)
4 ½-W resistors: 1 kΩ, 2.2 kΩ, 3.6 kΩ, 10 kΩ
1 VOM (analog or digital multimeter)

PROCEDURE

Negative Power Supply

1. In Fig. 19-1, calculate V_B, V_E, and V_C. Record your answers in Table 19-1.
2. Connect the circuit of Fig. 19-1. Measure and record the quantities listed in Table 19-1.
3. Repeat Steps 1 and 2 for the other transistors.

Positive Power Supply

4. In Fig. 19-2, calculate V_B, V_E, and V_C. Record your answers in Table 19-2.
5. Connect the circuit of Fig. 19-2. Measure and record the quantities of Table 19-2.
6. Repeat Steps 4 and 5 for the other transistors.

Troubleshooting (Optional)

7. In Fig. 19-2, assume R_1 is open. Estimate and record all voltages listed in Table 19-3.
8. Repeat Step 7 for the other troubles listed in Table 19-3.
9. Connect the circuit of Fig. 19-2 with each trouble listed in Table 19-3. Measure and record all voltages.
10. Ask the instructor to insert a trouble in your circuit.
11. Locate the trouble logically as follows. Measure V_B, V_E, and V_C. Then look at Fig. 19-2, visualizing

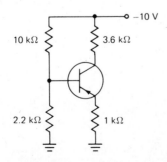

Figure 19-1

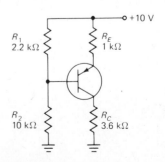

Figure 19-2

these voltages. Try to figure out which trouble would produce these voltages. When you think you have figured out the trouble, confirm your answer by consulting Table 19-3.

12. Repair the trouble and check that the circuit is working correctly.

13. Repeat Steps 10 to 12 as often as the instructor indicates.

Design (Optional)

14. Design an LED driver like Fig. 19-3 with a stiff voltage divider to meet the following specifications: $V_{CC} = 5$ V and $I_C = 20$ mA. You may assume a typical h_{FE} of 200. Record your design values (nearest standard resistances) in Table 19-4. Calculate and record I_C for your design.

15. Connect your design. Measure and record I_C. Repeat this measurement for the other transistors.

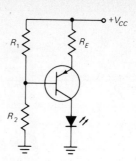

Figure 19-3

Computer (Optional)

16. Write and run a program that calculates the I_C and V_C for upside-down *pnp* bias like Fig. 19-2. Use INPUT statements to enter the required data.

DATA FOR EXPERIMENT 19

Table 19-1. Negative Power Supply

Transistor	Calculated V_B	V_E	V_C	Measured V_B	V_E	V_C
1						
2						
3						

Table 19-2. Positive Power Supply

Transistor	Calculated V_B	V_E	V_C	Measured V_B	V_E	V_C
1						
2						
3						

Table 19-3. Troubleshooting

Transistor	Estimated V_B	V_E	V_C	Measured V_B	V_E	V_C
Open R_1						
Open R_2						
Shorted R_2						
Open R_C						
Shorted R_C						
Open R_E						
Shorted R_E						
Open CE						
Shorted CE						

Table 19-4. Design

Values: $R_1 = $ _____; $R_2 = $ _____; $R_E = $ _____

Transistor	Calculated I_C	Measured I_C
1		
2		
3		

QUESTIONS FOR EXPERIMENT 19

1. Ideally, the voltage divider of Fig. 19-1 produces which of the following base   (   )
   voltages?
   (a) 0 V;   (b) −1.1 V;   (c) −1.8 V;   (d) −6.03 V.
2. The measured emitter voltage of Fig. 19-1 was closest to: (   )
   (a) 0 V;   (b) −1.1 V;   (c) −1.8 V;   (d) −6.03 V.
3. The measured collector voltage of Fig. 19-2 was closest to: (   )
   (a) 0 V;   (b) 3.97 V;   (c) 8.2 V;   (d) 8.9 V.
4. The emitter voltage measured in Fig. 19-2 was closest to: (   )
   (a) 0 V;   (b) slightly positive;   (c) slightly negative;   (d) 8.9 V.
5. With upside-down *pnp* bias, the emitter voltage is approximately: (   )
   (a) 0.7 V less than V_B;   (b) 0.7 V more than V_B;   (c) unknown;   (d) less
   than V_C.
6. What is the direction of current through each component of Fig. 19-2? Your answers
   should be up, down, left, or right for each resistor and transistor.

Troubleshooting (Optional)

7. Name all the troubles you found that produced a collector voltage of 0 V.

8. In Fig. 19-2, suppose $V_B = 10$ V, $V_E = 10$ V, and $V_C = 0$. What is the trouble?

Design (Optional)

9. Explain how you arrived at the value of R_E in your design.

10. Optional. Instructor's question.

EXPERIMENT 20

TRANSISTOR BIAS

▼

Before we can use a transistor to amplify an ac signal, we have to set up a quiescent (Q) point of operation, typically near the middle of the dc load line. Then, the incoming ac signal can produce fluctuations above and below this Q point. The three most primitive forms of bias are base bias, emitter-feedback bias, and collector-feedback bias. As you know, these are not the best ways to bias a transistor if you want a stable Q point. Nevertheless, you may occasionally see these biasing methods used with small-signal amplifiers. In this experiment you will connect all three types of bias to verify the operation as discussed in your textbook.

As discussed in Experiment 14, the most common transistor troubles are the collector-emitter short and collecter-emitter open. We will simulate the collector-emitter short by putting a jumper between the collector, base, and emitter; this is equivalent to shorting both diodes. We will simulate a collector-emitter open by removing the transistor from the circuit; this is equivalent to opening both diodes.

REQUIRED READING

Chapter 8 (Secs. 8-6, 8-9, and 8-10) of *Electronic Principles*, 5th ed.

EQUIPMENT

1 power supply: 15 V
3 transistors: 2N3904 (or equivalent)
7 ½-W resistors: 100 Ω, 750 Ω, 910 Ω, 1 kΩ, 200 kΩ, 270 kΩ, 430 kΩ

1 VOM (analog or digital multimeter)

PROCEDURE

Base Bias

1. Refer to the data sheet of a 2N3904 in the Appendix. Notice that the dc current gain h_{FE} has minimum value of 100 and a maximum value of 300 for an I_C of 10 mA. The typical value is not listed. For this experiment we will assume the typical value is 200.

2. In Fig. 20-1, use the typical h_{FE} to calculate I_B, I_C, and V_C. Record your answers in Table 20-1.

3. Connect the circuit of Fig. 20-1. Measure and record the quantities listed in Table 20-1.

4. Repeat Steps 2 and 3 for the other transistors.

Emitter-Feedback Bias

5. In Fig. 20-2, use the typical h_{FE} to calculate I_C, V_C, and V_E. Record your answers in Table 20-2.

6. Connect the emitter-feedback bias of Fig. 20-2. Measure and record the quantities of Table 20-2.

7. Repeat Steps 5 and 6 for the other transistors.

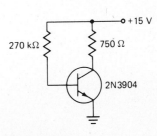

Figure 20-1

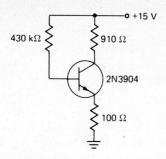

Figure 20-2

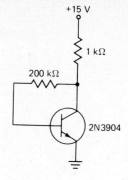

Figure 20-3

Collector-Feedback Bias

8. In Fig. 20-3, use the typical h_{FE} to calculate and record the quantities of Table 20-3.
9. Connect the circuit of Fig. 20-3. Measure and record all quantities listed in Table 20-3.
10. Repeat Steps 8 and 9 for the other transistors.

Troubleshooting (Optional)

11. In Fig. 20-3, assume the base resistor is open. Estimate and record the collector voltage V_C in Table 20-4.
12. Repeat Step 11 for the other troubles listed in Table 20-4.
13. Connect the circuit of Fig. 20-3 with each trouble listed in Table 20-4. Measure and record the collector voltage.

Design (Optional)

14. Design a collector-feedback biased circuit with a 2N3904 to meet the following specifications: $V_{CC} =$ 10 V and $I_C = 2$ mA. Hint: Look at Fig. 15 of the 2N3904 data sheet. Calculate and record the quantities listed in Table 20-5.
15. Connect your design. Measure and record the quantities of Table 20-5.

Computer (Optional)

16. Enter and run this program:

```
10 PRINT "1. BASE BIAS"
20 PRINT "2. EMITTER-FEEDBACK BIAS"
30 PRINT "3. COLLECTOR-FEEDBACK
   BIAS"
40 PRINT: PRINT "ENTER CHOICE": IN-
   PUT C
50 ON C GOTO 1000, 2000, 3000
60 END
1000 PRINT "YOU SELECTED BASE BIAS":
   END
2000 PRINT "YOU SELECTED EMITTER-
   FEEDBACK BIAS": END
3000 PRINT "YOU SELECTED COLLECTOR-
   FEEDBACK BIAS": END
```

17. Write and run a program that calculates the collector current for base bias, emitter-feedback bias, and collector-feedback bias. Use a menu like Step 16 and include INPUT statements to enter the required data.

DATA FOR EXPERIMENT 20

Table 20-1. Base Bias

Transistor	I_B	Calculated I_C	V_C	I_B	Measured I_C	V_C
1						
2						
3						

Table 20-2. Emitter-Feedback Bias

Transistor	I_C	Calculated V_C	V_E	I_C	Measured V_C	V_E
1						
2						
3						

Table 20-3. Collector-Feedback Bias

Transistor	V_B	Calculated I_C	V_C	V_B	Measured I_C	V_C
1						
2						
3						

Table 20-4. Troubleshooting

Trouble	Estimated V_C	Measured V_C
Open 200 kΩ		
Shorted 200 kΩ		
Open 1 kΩ		
Shorted 1 kΩ		
Open collector-emitter		
Shorted collector-emitter		

Table 20-5. Design: R_B = _____; R_C = _____

Transistor	Calculated V_C	Measured V_C
1		
2		
3		

QUESTIONS FOR EXPERIMENT 20

1. Base bias has an unstable Q point because of the variation in: ( )
   (a) base current: (b) V_{BE}; (c) base resistance; (d) h_{FE}.
2. When the collector current increases in a base-biased circuit, the collector ( )
   voltage:
   (a) increases; (b) stays the same; (c) decreases.
3. In Fig. 20-2, the collector saturation current has a value of approximately: ( )
   (a) 5 mA; (b) 10 mA; (c) 15 mA; (d) 20 mA.
4. The measured data of Table 20-3 show that the V_{BE} drop was closest to: ( )
   (a) 0; (b) 0.3 V; (c) 0.7 V; (d) 7.85 V.
5. Of the three circuits tested, which had the most stable Q point? ( )
   (a) base bias; (b) emitter-feedback bias; (c) collector-feedback
   bias; (d) voltage-divider bias.
6. Briefly discuss the Q point for the three circuits tested.

Troubleshooting (Optional)

7. Assume you are troubleshooting a circuit like Fig. 20-3. If you measure a collector
   voltage V_C of 15 V, what are three possible troubles?

8. Name two possible troubles in Fig. 20-3 that would produce a collector voltage of
   zero.

Design (Optional)

9. Explain how you calculated your design values in Table 20-5.

10. Optional. Instructor's question.

EXPERIMENT 21

COUPLING AND BYPASS CAPACITORS

▼

Capacitive reactance decreases as frequency increases. Because of this, a capacitor has a large impedance at low frequencies and a small impedance at high frequencies. As an approximation, we can say that a capacitor is a dc open and an ac short. When used in amplifiers, capacitors can couple the signal from one active node to another, or they can bypass the signal from an active node to ground.

REQUIRED READING

Chapter 9 (Secs. 9-1 and 9-2) of *Electronic Principles*, 5th ed.

EQUIPMENT

1 audio generator
4 ½-W resistors: two 22 kΩ, 68 kΩ, 100 kΩ
1 capacitor: 0.022 μF
1 oscilloscope

PROCEDURE

1. Calculate the critical frequency in Fig. 21-1a. Fill in the values of f_c, $10f_c$, and $0.1f_c$ in Table 21-1 under f.
2. Connect the circuit of Fig. 21-1a.
3. Adjust the audio generator to get a frequency of f_c and an input voltage v_{in} of 1 V peak-to-peak on the oscilloscope.
4. Measure the output voltage v_{out} and record in Table 21-1.
5. Change the frequency to $10f_c$ and readjust to get a v_{in} of 1 V p-p. Measure v_{out} and record in Table 21-1.
6. Change the frequency to $0.1f_c$ and check that v_{in} is 1 V p-p. Measure and record v_{out}.
7. Calculate the critical frequency in Fig. 21-1b. Fill in the values of f_c, $10f_c$, and $0.1f_c$ in Table 21-2.

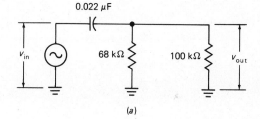

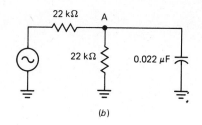

Figure 21-1

8. Connect the circuit of Fig. 21-1b without the capacitor. Adjust the frequency to f_c.
9. Set the signal level to 1 V p-p across the lower 22-kΩ resistor.
10. Connect the capacitor between point A and ground. Then measure and record v_A.
11. Remove the capacitor and change the frequency to $10f_c$. Then repeat Steps 9 and 10.
12. Remove the capacitor and change the frequency to $0.1f_c$. Then repeat Steps 9 and 10.

DATA FOR EXPERIMENT 21

Table 21-1. Coupling Capacitor

f	v_{out}
f_c	
$10f_c$	
$0.1f_c$	

Table 21-2. Bypass Capacitor

f	v_A
f_c	
$10f_c$	
$0.1f_c$	

QUESTIONS FOR EXPERIMENT 21

1. A coupling capacitor ideally looks like a dc: ( )
   (a) open and ac open; (b) open and ac short; (c) short and ac open;
   (d) short and ac short.
2. A small Thevenin resistance means the bypass capacitor must be: ( )
   (a) small; (b) large; (c) unaffected; (d) open.
3. The value of f_c in Table 21-1 is closest to: ( )
   (a) 100 Hz; (b) 180 Hz; (c) 1 kHz; (d) 5 kHz.
4. The value of f_c in Table 21-2 is closest to: ( )
   (a) 100 Hz; (b) 500 Hz; (c) 650 Hz; (d) 1 kHz.
5. If we want an f_c of 18 Hz in Fig. 21-1a, we have to change the capacitor to ( )
   approximately:
   (a) 0.1 μF; (b) 0.2 μF; (c) 0.5 μF; (d) 1 μF.
6. In Table 21-1, the output voltage at f_c is closest to: ( )
   (a) 0.707 V; (b) 0.9 V; (c) 0.99 V; (d) 1 V.
7. In Table 21-2, the output voltage at f_c is closest to: ( )
   (a) 0.01 V; (b) 0.15 V; (c) 0.707 V; (d) 1 V.
8. When the input voltage is 1 V in Fig. 21-1a, the output voltage at $10f_c$ is ( )
   closest to:
   (a) 0; (b) 0.707 V; (c) 0.9 V; (d) 1 V.
9. Optional. Instructor's question.

10. Optional. Instructor's question.

EXPERIMENT 22

AC EMITTER RESISTANCE

▼

The ac emitter resistance equals the change in base-emitter voltage divided by the change in emitter current. Because the graph of emitter current versus base-emitter voltage is exponential, it is possible to derive the following relation for the ac emitter resistance: $r'_e = 25$ mV/I_E. In this experiment, you will connect a circuit that allows you to verify the foregoing relation.

REQUIRED READING

Chapter 9 (Secs. 9-3 to 9-5) of *Electronic Principles*, 5th ed.

EQUIPMENT

1 audio generator

2 power supplies: 9 V and adjustable from at least 1 to 12 V
3 transistors: 2N3904 (or almost any small-signal *npn* silicon transistor)
3 ½-W resistors: 1 kΩ, two 10 kΩ
1 capacitor: 0.1 μF
1 oscilloscope
1 ac millivoltmeter

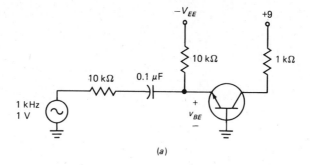

(a)

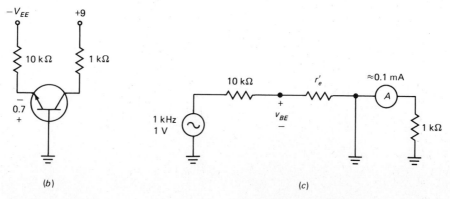

(b)

(c)

Figure 22-1

PROCEDURE

1. Figure 22-1a shows a circuit that measures r'_e. In the dc equivalent circuit of Fig. 22-1b, the dc emitter current equals the voltage across the 10kΩ resistor divided by 10 kΩ. Calculate the value of I_E for each value of $-V_{EE}$ shown in Table 22-1. Record the values of I_E.

2. Using Eq. (9-9), calculate and record the value of r'_e for each I_E.

3. Connect the circuit of Fig. 22-1a. Set $-V_{EE}$ at -10.7 V. Adjust the audio generator to 1 kHz. With the ac millivoltmeter in parallel with the audio generator, adjust the signal level to 1 V rms.

4. Once the audio generator has been set to 1 V rms, the approximate ac equivalent circuit looks like Fig. 22-1c. Notice most of the 1 V will be across the 10-kΩ resistance because r'_e is very small by comparison (around 25 Ω). This is why the alternating current i_e is approximately 0.1 mA.

5. Connect the ac millivoltmeter and the oscilloscope to measure v_{be}, the ac voltage between the emitter and ground in Fig. 22-1a.

6. The oscilloscope should display a small sine wave. If so, read the value of v_{be} with the ac millivoltmeter. Record in Table 22-2.

7. Reduce the $-V_{EE}$ supply to -5.7 V. The sine wave on the oscilloscope should increase. Read the new value of v_{be} on the ac millivoltmeter and record in Table 22-2.

8. Reduce the $-V_{EE}$ supply to -3.2 V. Read the new value of v_{be} and record.

9. In Fig. 22-1c, r'_e is much smaller than 10 kΩ. Therefore, almost all of the 1 V is across the 10 kΩ. For this reason, the ac emitter current i_e is approximately equal to 0.1 mA. Now, you can calculate r'_e by dividing each v_{be} by 0.1 mA. Calculate and record the values in Table 22-2.

DATA FOR EXPERIMENT 22

Table 22-1 Calculations

$-V_{EE}$	I_E	r'_e
-10.7		
-5.7		
-3.2		

Table 22-2 Measurements

$-V_{EE}$	v_{be}	r'_e
-10.7		
-5.7		
-3.2		

QUESTIONS FOR EXPERIMENT 22

1. In Table 22-1, each time emitter current is cut in half, ac emitter resistance: ( )
   (a) goes down by a factor of two; (b) doubles; (c) quadruples;
   (d) stays the same.
2. In Fig. 22-1a, the positive end of the $-V_{EE}$ supply is: ( )
   (a) grounded; (b) connected to the 10-kΩ resistor; (c) connected to the emitter.
3. When $-V_{EE}$ equals -10.7 V in Fig. 22-1b, V_{CE} equals: ( )
   (a) 0.7 V; (b) 1 V; (c) 5 V; (d) 8 V.
4. The value of r'_e equals: ( )
   (a) v_{be}/i_b; (b) v_{be}/i_e; (c) v_{ce}/i_e; (d) v_{ce}/i_b.
5. Because r'_e is only a small part of the resistance in Fig. 22-1c, the circuit ( )
   driving r'_e is approximately an ideal:
   (a) voltage source; (b) current source; (c) resistance; (d) capacitor.
6. According to Tables 22-1 and 22-2, the value of v_{be} for an I_E of 0.5 mA is ( )
   closest to:
   (a) 2.5 mV; (b) 5 mV; (c) 50 mV; (d) 250 mV.
7. Table 22-2 confirms that r'_e is: ( )
   (a) 25 Ω; (b) equal to v_{be}/i_b; (c) inversely proportional to v_{be};
   (d) inversely proportional to I_E.
8. In Fig. 22-1c, r'_e is much smaller than: ( )
   (a) 25 Ω; (b) 50 Ω; (c) 100 Ω; (d) 10 kΩ.

9. Optional. Instructor's question.

10. Optional. Instructor's question.

EXPERIMENT 23

THE CE AMPLIFIER

▼

After the transistor of a CE amplifier has been biased with its *Q* point near the middle of the dc load line, you can couple a small ac signal into the base. This produces an amplified ac signal at the collector. In this experiment you will build a CE amplifier and measure its voltage gain, as well as looking at the dc and ac waveforms throughout the circuit.

REQUIRED READING

Chapter 10 (Secs. 10-1 to 10-4) of *Electronic Principles*, 5th ed.

EQUIPMENT

1 audio generator
1 power supply: 10 V
3 transistors: 2N3904 (or equivalent)
4 ½-W resistors: 1 kΩ, 2.2 kΩ, 3.6 kΩ, 10 kΩ
2 capacitors: 1 μF, 470 μF (10-V rating or better)
1 oscilloscope

PROCEDURE

DC and AC Voltages

1. In Fig. 23-1, calculate the dc voltage at the base, emitter, and collector. Record your answers in Table 23-1.

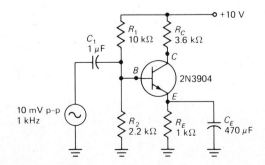

Figure 23-1

2. Calculate and record the peak-to-peak ac voltages at the base, emitter, and collector.
3. Connect the circuit. Adjust the signal generator to get an input signal of 10 mV peak-to-peak at 1 kHz.
4. Look at the base, emitter, and collector. At each point, use the oscilloscope on dc input to measure dc voltage and on ac input to measure the ac peak-to-peak voltage. Record all voltages in Table 23-1.
5. With the oscilloscope on dc input, you should see waveforms like those of Fig. 9-15 in your textbook. This confirms that the total voltages are the sum of dc and ac components.

Phase Inversion

6. If you are using a dual-trace or dual-beam oscilloscope, look at the base signal with one input and the collector signal with the other input. Also use the collector signal to drive the external trigger of the oscilloscope. (If in doubt about how to use the external trigger, ask the instructor.) Notice that the collector signal is 180° out of phase with the base signal.
7. If you are using a single-trace oscilloscope, externally trigger the oscilloscope with the collector signal. (If in doubt, ask the instructor about this.) Look first at the base signal, then at the collector signal. Notice that the signals are 180° out of phase.

Voltage Gain

8. In Fig. 23-1, use Eq. (9-9) in your textbook to calculate the ideal emitter resistance r'_e. Use Eq. (10-

9) to calculate the voltage gain A. Record your answers in Table 23-2.

9. Connect the circuit with any of the three transistors. Measure and record the input and output ac voltages.

10. Calculate the actual voltage gain using the v_{out} and v_{in} measured in Step 9. Next calculate r'_e using the ratio R_C/A. Record your experimental A and r'_e in Table 23-2.

11. Repeat Steps 8 to 10 for the other transistors.

Troubleshooting (Optional)

12. In Fig. 23-1, assume C_1 is open. Estimate the peak-to-peak ac voltage at the base, emitter, and collector. Record in Table 23-3.

13. Repeat Step 12 for each trouble listed in Table 23-3.

14. Connect the circuit with each trouble. Measure and record the ac voltages.

Design (Optional)

15. Select a value of collector resistance in Fig. 23-1 to produce a theoretical voltage gain of 100. Using the nearest standard resistance, calculate and record the quantities of Table 23-4.

16. Connect the circuit with your design value of R_C. Measure and record the quantities listed in Table 23-4.

17. Repeat Step 16 for the other transistors.

Computer (Optional)

18. Enter and run this program:

```
10 REM VOLTAGE GAIN
20 PRINT "ENTER COLLECTOR RESIS-
   TANCE": INPUT RC
30 PRINT "ENTER R PRIME E": INPUT RPE
40 A = RC/RPE
50 PRINT "THE VOLTAGE GAIN EQUALS":
   PRINT A
60 END
```

19. Write and run a program that calculates the voltage gain for any circuit like Fig. 23-1 using the ideal equation: $A = R_C/r'_e$. Use INPUT statements to enter R_1, R_2, R_E, and R_C. The program should then calculate the theoretical value of r'_e and A.

DATA FOR EXPERIMENT 23

Table 23-1. CE Amplifier

	Calculated			Measured		
	B	E	C	B	E	C
dc						
ac						

Table 23-2. Voltage Gain

Transistor	Calculated		Measured		Experimental	
	r'_e	A	v_{in}	v_{out}	A	r'_e
1						
2						
3						

Table 23-3. Troubleshooting

Trouble	Estimated			Measured		
	v_b	v_e	v_c	v_b	v_e	v_c
Open C_1						
Open R_2						
Open R_E						

Table 23-4. Design

Transistor	Calculated			Measured		
	r'_e	R_C	A	v_{in}	v_{out}	A
1						
2						
3						

QUESTIONS FOR EXPERIMENT 23

1. The CE amplifier of Fig. 23-1 has a theoretical r'_e of: ( )
   (a) 22.7 Ω; (b) 1 kΩ; (c) 3.6 kΩ; (d) 10 kΩ.
2. Ideally, the CE amplifier of Fig. 23-1 has a voltage gain of approximately: ( )
   (a) 1; (b) 3.6; (c) 4.54; (d) 159.
3. The emitter of Fig. 23-1 had little or no ac signal because of: ( )
   (a) the emitter resistor; (b) the input coupling capacitor; (c) the emitter bypass capacitor; (d) the weak base signal.

4. The voltage at collector was closest to: ( )
(a) 6 V dc and 10 mV ac; (b) 1.8 V dc and 1.6 V ac; (c) 1.1 V dc and 10 mV ac; (d) 6 V dc and 1.6 V ac.

5. The dc bias of the transistor is undisturbed by the dc resistance of the signal ( ) generator because the input coupling capacitor:
(a) blocks dc; (b) transmits ac; (c) blocks ac; (d) transmits dc.

6. As briefly as possible, explain how the circuit of this experiment amplifies the signal.

Troubleshooting (Optional)

7. What happens to the dc voltages of the circuit when the coupling capacitor is open? What happens to the ac voltages?

8. Explain the measured voltages you got with an open R_E.

Design (Optional)

9. Explain how you selected the value of load resistance.

10. Optional. Instructor's question.

EXPERIMENT 24

OTHER CE AMPLIFIERS

▼

Because of the input impedance of a CE amplifier, some of the ac signal may be dropped across the source impedance. Furthermore, the Thevenin equivalent of the amplifier output is an ac generator in series with the output impedance of the amplifier. When a load resistance is connected to the amplifier, some of the ac signal is dropped across the output impedance.

One way to increase the input impedance is to use a swamping resistor in the emitter circuit. This also stabilizes the voltage gain against changes in r'_e. Because the swamping resistor reduces the voltage gain, it may be necessary to cascade two swamped amplifiers to get the same voltage gain as a single unswamped stage.

REQUIRED READING

Chapter 10 (Secs. 10-1 to 10-5) of *Electronic Principles*, 5th ed.

EQUIPMENT

1 audio generator
1 power supply: 10 V
3 transistors: 2N3904 (or equivalent)
6 ½-W resistors: two 1 kΩ, 1.5 kΩ, 2.2 kΩ, 3.6 kΩ, 10 kΩ
3 capacitors: two 1 μF, 470 μF (10-V rating or better)
1 oscilloscope

PROCEDURE

CE Amplifier with Source and Load Resistances

1. In Fig. 24-1, assume h_{fe} (same as β) is 150. Calculate the input impedance of the stage. Also calculate the peak-to-peak base voltage and the peak-to-peak collector voltage. Record your answers in Table 24-1.
2. Connect the circuit. Adjust the signal generator to get a source signal of 20 mV peak-to-peak at 1 kHz.

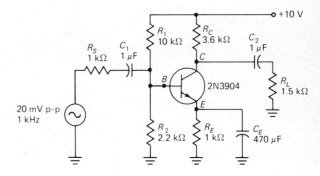

Figure 24-1

(Measure this between the left end of the source resistance and ground.)

3. Look at the base and collector. At each point, use the oscilloscope on dc input to verify that the waveforms are the sum of dc and ac components. Also look at the emitter. Because of the bypass capacitor, the emitter should have only a dc component.
4. With the oscilloscope on ac input, measure the peak-to-peak voltage at the base and collector. Record your data in Table 24-1.
5. Repeat Step 4 for the two other transistors.

Swamped Amplifier

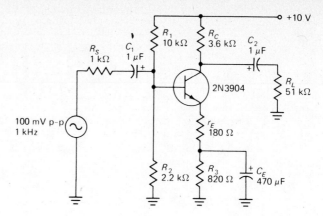

Figure 24-2

6. In Fig. 24-2, assume an h_{fe} of 150 and calculate the input impedance of the stage. Also calculate the peak-to-peak voltage at the base and collector. Record your answers in Table 24-2.

7. Connect the circuit. Measure and record the peak-to-peak voltage at the base and collector.

8. Repeat Steps 6 and 7 for the other two transistors.

Troubleshooting (Optional)

9. In Fig. 24-2, assume C_E is open. Estimate the peak-to-peak ac voltage at the base, emitter, and collector. Record in Table 24-3.

10. Repeat Step 9 for each trouble listed in Table 24-3.

11. Connect the circuit with each trouble. Measure and record the ac voltages.

Design (Optional)

12. Select a value of swamping resistance in Fig. 24-2 to produce an unloaded voltage gain of 10 from the base to the collector. Using the nearest standard resistance, calculate and record the quantities of Table 24-4.

13. Connect the circuit with your design value of r_E. Measure and record the quantities listed in Table 24-4.

14. Repeat Step 13 for the other transistors.

Computer (Optional)

15. Write and run a program that calculates the voltage gain for a circuit like Fig. 24-1. Use INPUT statements to enter R_1, R_2, R_E, R_C, R_S, R_L, and h_{fe}.

DATA FOR EXPERIMENT 24

Table 24-1. CE Amplifier with Source and Load Resistances

		Calculated		Measured	
Transistor	Z_{in}	v_b	v_c	v_b	v_c
1					
2					
3					

Table 24-2. Swamped Amplifier

		Calculated		Measured	
Transistor	Z_{in}	v_b	v_c	v_b	v_c
1					
2					
3					

Table 24-3. Troubleshooting

	Estimated			Measured		
Trouble	v_b	v_e	v_c	v_b	v_e	v_c
Open C_E						
Shorted C_E						
Open collector-emitter						
Shorted collector-emitter						
Open C_2						
Shorted C_2						

Table 24-4. Design

		Calculated		Measured	
Transistor	r_e	v_b	v_c	v_b	v_c
1					
2					
3					

QUESTIONS FOR EXPERIMENT 24

1. The calculated base voltage in Table 24-1 is approximately: (   )
   (a) 10.8 mV;   (b) 20 mV;   (c) 250 mV;   (d) 500 mV.

2. The calculated collector voltage in Table 24-1 was closest to: ( )
   (a) 10.8 mV;   (b) 20 mV;   (c) 250 mV;   (d) 500 mV.

3. In Table 24-2, the measured base voltage was closest to: ( )
   (a) 12 mV;   (b) 20 mV;   (c) 63 mV;   (d) 1 V.

4. The voltage gain from base to collector in the swamped amplifier was closest ( )
   to:
   (a) 1;   (b) 5;   (c) 10;   (d) 15.

5. Compared with the CE amplifier, the swamped amplifier had a: ( )
   (a) lower input impedance;   (b) higher output impedance;   (c) lower voltage gain;   (d) lower ac collector voltage.

6. Explain why an amplifier with a swamping resistor has a more stable voltage gain than one without.

Troubleshooting (Optional)

7. What happens to the voltage gain of an amplifier when the emitter bypass capacitor is open? Explain why this happens.

8. Explain what happens when the emitter bypass capacitor is shorted.

Design (Optional)

9. How did you arrive at the value of the swamping resistor?

10. Optional. Instructor's question.

EXPERIMENT 25

CASCADED CE STAGES

▼

The amplified signal out of a CE stage can be used as the input to another CE stage. In this way, we can build a multistage amplifier with very large voltage gain. Because a CE stage has an input impedance, there is a loading effect on the preceding stage. In other words, the loaded voltage gain is less than the unloaded voltage gain. In this experiment, you will build a two-stage amplifier using swamped stages to stabilize the overall voltage gain.

REQUIRED READING

Chapter 10 (Sec. 10-6) of *Electronic Principles*, 5th ed.

EQUIPMENT

1 audio generator
1 power supply: 10 V
2 transistors: 2N3904 (or equivalent)
12 ½-W resistors: two 68 Ω, three 1 kΩ, one 1.2 kΩ, two 2.2 kΩ, two 3.6 kΩ, two 10 kΩ
5 capacitors: three 1 μF, two 47 μF (10-V rating or better)
1 VOM (analog or digital multimeter)
1 oscilloscope

PROCEDURE

Calculations

1. In Fig. 25-1, calculate the dc voltages at the base, emitter, and collector of each stage. Record your answers in Table 25-1.
2. Next, examine Fig. 11 of the 2N3904 data sheet in the Appendix. Read the typical value of h_{fe}. Record this value at the top of Table 25-2.
3. Calculate the peak-to-peak ac voltage at the base, emitter, and collector of each stage (Fig. 25-1) using the h_{fe} of Step 2. Record all ac voltages in Table 25-2.

Tests

4. Connect the two-stage amplifier of Fig. 25-1.
5. Measure the dc voltage at the base, emitter, and collector of each stage. Record your data in Table 25-1. Within the tolerance of the resistors being used, the measured voltages should agree with your calculated voltages.
6. Measure the peak-to-peak ac voltage at the base, emitter, and collector of each stage. Record your data in Table 25-2. These measured ac voltages should agree with your calculated values.

Loading Effects

7. Open the coupling capacitor between the first and second stage. Look at the ac voltage on the first collector. Reconnect the coupling capacitor and notice that the ac signal decreases significantly. Do not continue until you understand and can explain why the signal decreases.
8. Open the coupling capacitor between the second stage and the load resistor. Look at the ac voltage on the second collector. Reconnect the coupling capacitor and notice the decrease in signal strength. Once more, you should be able to explain why this happens.

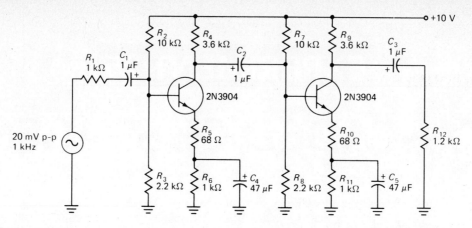

· **Figure 25-1**

Troubleshooting (Optional)

9. In Fig. 25-1, assume C_4 is open. Does this produce a trouble in the first or second stage? Record your answer (1 or 2) in Table 25-3.

10. Insert the foregoing trouble in your circuit. You will be measuring dc and ac voltages. Before you measure each voltage, make a ballpark estimate of its value. Then when you measure the voltage, you will know whether or not it is a clue to the trouble.

11. Estimate each voltage in Table 25-3 for the stage with the trouble. Measure and record the voltage.

12. Repeat Steps 9 to 11 for each of the troubles listed in Table 25-3.

13. Ask the instructor to insert a trouble in your circuit.

14. Locate, repair, and fix the trouble.

15. Repeat Steps 13 and 14 as often as indicated by the instructor.

Design (Optional)

16. Select a swamping resistor for the second stage to get an overall voltage gain of approximately 75. Record your design value at the top of Table 25-4.

17. Change the swamping resistor to your design value. Measure and record the voltage gain of the first stage (base to collector). Then measure and record the voltage gain of the second stage.

18. Measure and record the overall voltage gain (first base to second collector).

Computer (Optional)

19. Write and run a program that calculates the voltage gain for a circuit like Fig. 25-1. Use INPUT statements to enter R_1 through R_{12}, h_{fe}, and any other data needed to solve the problem.

DATA FOR EXPERIMENT 25

Table 25-1. DC Voltages

Stage	Calculated V_B	V_E	V_C	Measured V_B	V_E	V_C
1						
2						

Table 25-2. AC Voltages: h_{fe} = _____

Stage	Calculated v_b	v_e	v_c	Measured v_b	v_e	v_c
1						
2						

Table 25-3. Troubleshooting

Trouble	Stage	DC Voltages V_B	V_E	V_C	AC Voltages v_b	v_e	v_c
Open C_4							
Shorted R_4							
Shorted R_{10}							
Open R_3							
Open C_5							

Table 25-4. Design: r_E = _____

A_1 = _____

A_2 = _____

A = _____

QUESTIONS FOR EXPERIMENT 25

1. The calculated dc base voltage of the first stage was approximately: ( )
   (a) 1.1 V; (b) 1.8 V; (c) 6.28 V; (d) 10 V.
2. The measured dc collector voltage of the second stage was closest to: ( )
   (a) 1.1 V; (b) 1.8 V; (c) 6.28 V; (d) 10 V.
3. The ac base voltage of the first stage was closest to: ( )
   (a) 5 mV; (b) 12 mV; (c) 100 mV; (d) 1.4 V.
4. The ac emitter voltage of the second stage was closest to: ( )
   (a) 5 mV; (b) 12 mV; (c) 100 mV; (d) 1.4 V.

5. The voltage gain from the base of the first stage to the collector of the second   (   )
   stage was closest to:
   (a) 10;   (b) 115;   (c) 230;   (d) 1000.

6. Explain why the signal decreased when the coupling capacitor was reconnected in
   Step 7.

Troubleshooting (Optional)

7. Explain what happens when an emitter bypass capacitor opens.

8. Suppose you get a collector-emitter short in the first stage of Fig. 25-1. What is the
   approximate input impedance looking into the base of the first stage? Explain your
   answer.

Design (Optional)

9. There is a simple way to modify Fig. 25-1 for use with *pnp* transistors. Explain what
   changes need to be made.

10. Optional. Instructor's question.

EXPERIMENT 26

AC LOAD LINES

▼

Every amplifier sees two loads: a dc load and an ac load. Because of this, every amplifier has two load lines: a dc load line and an ac load line. In this experiment you will build a circuit that displays the ac load line on an oscilloscope.

REQUIRED READING

Chapter 11 (Secs. 11-1 and 11-2) of *Electronic Principles*, 5th ed.

EQUIPMENT

1 audio generator
1 power supply: 9 V
1 transistor: 2N3904 (or almost any small-signal *npn* silicon transistor)
5 ½-W resistors: 100 Ω, 220 Ω, 4.7 kΩ, 10 kΩ, 1 MΩ
1 decade resistance box (if unavailable, use 50-kΩ potentiometer)
2 capacitors: 1 μF, 10 μF (10-V rating or better)
1 oscilloscope

PROCEDURE

1. Set up the oscilloscope as follows: Turn the horizontal sensitivity control to 1 V/cm (dc input) and the vertical sensitivity control to 100 mV/cm (dc input). Place the center spot in upper left-hand corner.
2. Connect the circuit of Fig. 26-1a. Notice the collector current passes through the 100-Ω resistor. In fact, each milliampere of collector current produces 100 mV of vertical input to the oscilloscope.
3. With the audio generator turned down to zero, the spot will be deflected. Since vertical sensitivity corresponds to 1 mA/cm and horizontal sensitivity to 1 V/cm, you can measure the values of I_{CQ} and V_{CEQ}. Record these values in Table 26-1.
4. Adjust the decade resistance box R to 2.5 kΩ. Be-

cause the 100-Ω resistor is small, the approximate ac equivalent circuit looks like Fig. 26-1b. Calculate and record the value of r_c in Table 26-1.
5. Increase the signal level out of the audio generator until the ac load line just hits the saturation and cutoff points. Measure and record the values of $I_{C(sat)}$ and $V_{CE(cutoff)}$ in Table 26-2.
6. Repeat Steps 4 and 5 for the two other values of R shown in Table 26-2.
7. For each ac load resistance of Table 26-2, calculate $I_{C(sat)}$ and $V_{CE(cutoff)}$ using Eqs. (11-16) and (11-17). Record these values in Table 26-3.

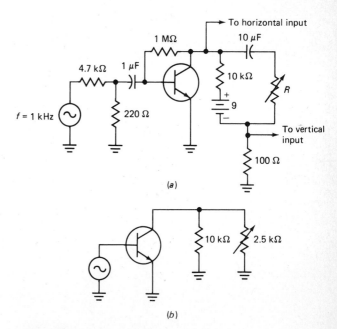

Figure 26-1

● ## DATA FOR EXPERIMENT 26

Table 26-1. Quiescent Values

I_{CQ}	
V_{CEQ}	

Table 26-2. AC Load Line

R	r_c	$I_{C(\text{sat})}$	$V_{CE(\text{cutoff})}$
2.5 kΩ			
10 kΩ			
1 MΩ			

Table 26-3. Calculations

AC Load	$I_{C(\text{sat})}$	$V_{CE(\text{cutoff})}$
1		
2		
3		

QUESTIONS FOR EXPERIMENT 26

1. In Fig. 26-1a, the current through the 100-Ω resistor is: ( )
   (a) direct current; (b) alternating current; (c) total current.
2. The reason the ac load line appears upside-down in this experiment is because ( )
   the:
   (a) transistor is *npn*; (b) voltage across the 100-Ω resistor is negative;
   (c) negative end of the 9-V supply is connected to the vertical input;
   (d) voltage across the 100-Ω resistor is positive.
3. When the audio generator is turned down to zero, the current through the ( )
   100-Ω resistor is identical to the:
   (a) ac collector current; (b) current through the 10-kΩ resistor;
   (c) current through the decade resistance; (d) base current.
4. The voltage to the vertical input equals the: ( )
   (a) current through the 100-Ω resistor; (b) collector voltage; (c) ac collector current; (d) product of total collector current and 100 Ω.
5. The voltage to the horizontal input equals the: ( )
   (a) base-emitter voltage; (b) collector-emitter voltage; (c) collector-base
   voltage; (d) vertical-input voltage.
6. As the ac load resistance increases, the ac load line becomes: ( )
   (a) straighter; (b) more vertical; (c) more horizontal.
7. When the value of R decreases in Fig. 26-1a, the ac load resistance: ( )
   (a) decreases; (b) increases; (c) stays the same; (d) impossible to say.

8. Optional. Instructor's question.

9. Optional. Instructor's question.

EXPERIMENT 27

CLASS A AMPLIFIERS

▼

In a class A amplifier, the transistor operates in the active region throughout the ac cycle. This is equivalent to saying the signal does not drive the transistor into either saturation or cutoff on the ac load line. With a CE amplifier, MPP is the smaller of $2I_{CQ}r_C$ or $2V_{CEQ}$. Some of the important quantities in a class A amplifier are the current drain, the maximum transistor power dissipation, the maximum unclipped load power, and the stage efficiency. In this experiment you will calculate and measure the voltages, currents and powers of a class A amplifier.

REQUIRED READING

Chapter 11 (Secs. 11-1 to 11-4) of *Electronic Principles*, 5th ed.

EQUIPMENT

1 audio generator
1 power supply: 15 V
1 transistor: 2N3904
7 ½-W resistors: 220 Ω, 1 kΩ, 1.5 kΩ, 1.8 kΩ, 2.2 kΩ, 3.6 kΩ, 10 kΩ
3 capacitors: two 1 μF, 470 μF (16-V rating or better)
1 VOM (analog or digital multimeter)
1 oscilloscope

PROCEDURE

CE Amplifier

1. In Fig. 27-1, calculate the quiescent collector current and the quiescent collector-emitter voltage. Record your answers in Table 27-1.
2. Calculate and record the MPP and the current drain of the stage.
3. Calculate the maximum transistor power dissipation, maximum unclipped load power, dc input power to the stage, and stage efficiency. Record your answers in the theoretical column of Table 27-2.
4. Connect the circuit. Reduce the signal generator to

zero. Use the VOM to measure I_{CQ} and V_{CEQ}. Record the data.

5. Use the oscilloscope to look at the load voltage. Adjust the signal generator until clipping starts on either half-cycle. Notice how the waveform appears squashed on the upper half-cycle and elongated on the lower half-cycle. This nonlinear distortion is being caused by the large changes in r'_e as the collector approaches cutoff and saturation.
6. Reduce the signal generator until the clipping stops. You have to estimate this as best you can because the clipping is soft as the operating point approaches cutoff. Back off enough from the clipping so that the upper peak appears rounded and smooth. Measure and record the peak-to-peak ac voltage. (This

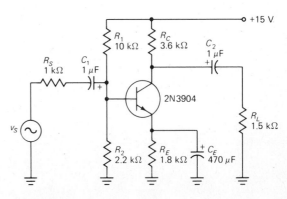

Figure 27-1

measured value is a rough approximation of the MPP value.

7. Measure and record the total current drain of the stage.
8. Calculate and record the experimental quantities listed in Table 27-2 using the measured data of Table 27-1.
9. Adjust the signal generator to get a peak-to-peak load voltage of 2 V. Notice how much nonlinear distortion there is.
10. Insert a swamping resistor of 220 Ω. Adjust the signal generator to get a load voltage of 2 V p-p and notice how the load signal appears less distorted than before. You should know from your textbook why this improvement occurs.

Troubleshooting (Optional)

11. In Fig. 27-1, assume R_2 is shorted. Calculate the MPP value and current drain for this trouble. Record in Table 27-3.
12. Repeat Step 11 for each trouble listed in Table 27-3.
13. Connect the circuit with each trouble. Measure and record MPP and I_S.

Design (Optional)

14. Select a value of R_E to get maximum MPP value in Fig. 27-1. Record the nearest standard resistance at the top of Table 27-4. Calculate and record the other quantities of Table 27-4.
15. Connect the circuit with your design value of R_E. Measure and record MPP and I_S in Table 27-4. Calculate and record the experimental quantities $P_{L(max)}$, P_S, and η using the measured data for MPP and I_S.

Computer (Optional)

16. Write and run a program for a CE amplifier that calculates the small-signal voltage gain, input impedance, and output impedance using H parameters. The inputs are h_{ie}, h_{fe}, h_{re}, h_{oe}, R_S, R_1, R_2, R_C, and R_L.
17. Write and run a program that prints out I_{CQ}, V_{CEQ}, MPP, I_S, $P_{D(max)}$, $P_{L(max)}$, P_S, and η for an amplifier like Fig. 27-1. The inputs are V_{CC}, R_1, R_2, R_E, R_C, and R_L.

DATA FOR EXPERIMENT 27

Table 27-1. CE Amplifier

	Calculated	Measured
I_{CQ}		
V_{CEQ}		
MPP		
I_S		

Table 27-2. Power and Efficiency

	Theoretical	Experimental
$P_{D(\max)}$		
$P_{L(\max)}$		
P_S		
η		

Table 27-3. Troubleshooting

	Estimated		Measured	
Trouble	MPP	I_S	MPP	I_S
Shorted R_2				
Open C_E				
Open R_L				
Collector-emitter open				

Table 27-4. Design: R_E = _____

	Theoretical	Experimental
MPP		
I_S		
$P_{L(\max)}$		
P_S		
η		

QUESTIONS FOR EXPERIMENT 27

1. The theoretical MPP value of Fig. 27-1 is approximately: ( )
   (a) 1.1V; (b) 2.35 V; (c) 9 V; (d) 15 V.

2. The total current drain of the amplifier was closest to: (   )
   (a) 1.1 mA;   (b) 2.3 mA;   (c) 4.8 mA;   (d) 6.9 mA.
3. The maximum transistor power dissipation of Fig. 27-1 is approximately: (   )
   (a) 0.46 mW;   (b) 10 mW;   (c) 35.1 mW;   (d) 50 mW.
4. Theoretically, the maximum efficiency of Fig. 27-1 is approximately: (   )
   (a) 0;   (b) 1.3%;   (c) 5%;   (d) 25%.
5. Inserting a swamping resistor in the circuit of Fig. 27-1:   (   )
   (a) reduced supply voltage: (b) increased quiescent collector current;
   (c) decreased nonlinear distortion;   (d) increased ac output compliance.
6. Explain why nonlinear distortion exists in a CE amplifier with a large output signal.

Troubleshooting (Optional)

7. What happens to the MPP value when the bypass capacitor CE opens? To the voltage gain?

8. Why did the MPP value increase when R_L was opened?

Design (Optional)

9. Explain how you selected the value of R_E to get maximum MPP.

10. Optional. Instructor's question.

EXPERIMENT 28

THE EMITTER FOLLOWER

▼

An emitter follower has high input impedance, low output impedance, and low nonlinear distortion. It is often used as a buffer stage between a high-impedance source and a low-resistance load. In this experiment you will build an emitter follower to verify its high input impedance and low output impedance.

REQUIRED READING

Chapter 12 (Secs. 12-1 to 12-4) of *Electronic Principles*, 5th ed.

EQUIPMENT

1 audio generator
1 power supply: 10 V
1 transistor: 2N3904 (or equivalent)
5 ½-W resistors: 51 Ω, 3.6 kΩ, 4.3 kΩ, two 10 kΩ
2 capacitors: 1 μF, 470 μF (10-V rating or better)
1 VOM (analog or digital multimeter)
1 oscilloscope

PROCEDURE

Emitter Follower

1. In Fig. 28-1, calculate the dc voltage at the base, emitter, and collector. Record your answers in Table 28-1.
2. Refer to the data sheet of 2N3904 to determine the typical h_{fe}. Record this value in Table 28-1.
3. Calculate and record the peak-to-peak ac voltage at the base, emitter, and collector.
4. Connect the circuit. Measure and record the dc voltage at the base, emitter, and collector.
5. Adjust the signal generator to get a source signal of 1 V peak-to-peak at 10 kHz. (Measure this between the left end of the source resistor and ground.)
6. Measure and record the peak-to-peak ac voltage at the base, emitter, and collector.

Output Impedance

7. Calculate the theoretical output impedance for the circuit of Fig. 28-1. Record your answer in Table 28-2.
8. Reduce the peak-to-peak source from 1 V to 100 mV.
9. Measure and record the peak-to-peak output voltage (unloaded).
10. Connect a load resistance of 51 Ω across the output.
11. Measure the loaded output voltage.
12. Calculate the output impedance of the emitter follower with the data of Steps 9 and 11. Record your experimental answer in Table 28-1.

Troubleshooting (Optional)

13. In Fig. 28-1, assume R_1 is open. Estimate the dc and ac voltages at the output emitter. Record your answers in Table 28-3.

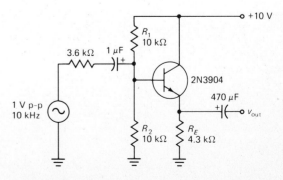

Figure 28-1

14. Repeat Step 13 for each trouble listed in Table 28-3.
15. Connect the circuit with each trouble. Measure and record the dc and ac voltages.

Design (Optional)

16. In Fig. 28-1, select a value of R_E to get a dc emitter current of 2.5 mA. Record your design value in Table 28-4.
17. Connect the circuit with your design value of R_E. Measure and record the dc emitter voltage. Measure and record and dc emitter current.

Computer (Optional)

18. Enter and run this program:

```
10  PRINT "ENTER VCC": INPUT VCC
20  PRINT "ENTER R1": INPUT R1
30  PRINT "ENTER R2": INPUT R2
40  K = R2/(R1 + R2)
50  VB = K * VCC: PRINT VB
60  STOP
70  VE = VB - 0.7: PRINT VE
80  STOP
90  GOTO 10
```

At each breakpoint, it will be necessary to enter CONT to continue.

19. Write and run a program that calculates the input and output impedance of an emitter follower like Fig. 28-1. The inputs are R_1, R_2, h_{fe}, r'_e, and R_E.

● ## DATA FOR EXPERIMENT 28

Table 28-1. Emitter Follower: h_{fe} = _____

	Calculated			Measured		
	B	E	C	B	E	C
dc						
ac						

Table 28-2. Output Impedance

Calculated r_{out} =

Unloaded v_{out} =

Loaded v_{out} =

Experimental r_{out} =

Table 28-3. Troubleshooting

	Estimated		Measured	
Trouble	V_E	v_e	V_E	v_e
Open R_1				
Shorted R_1				
Open R_2				
Shorted R_2				
Open R_E				
Shorted R_E				

Table 28-4. Design

Calculated R_E =

Measured V_E =

Measured I_E =

QUESTIONS FOR EXPERIMENT 28

1. The data of Table 28-1 show that the voltage gain of the emitter follower was   (   )
   approximately:
   (a) 0;   (b) 1;   (c) 4.3 V;   (d) 10 V.

2. The ac collector voltage of the emitter follower was closest to: (   )
   (a) 0;   (b) 0.58 V;   (c) 1 V;   (d) 10 V.

3. Because the ac emitter voltage approximately equals the ac base voltage in   (   )
Table 28-1, the input impedance of the base must be:
(**a**) 0;   (**b**) very low;   (**c**) 10 V;   (**d**) very high.

4. The calculated output impedance in Table 28-2 is closest to: (   )
(**a**) 1 Ω;   (**b**) 23 Ω;   (**c**) 42.4 Ω;   (**d**) 51 Ω.

5. The unloaded output voltage in Table 28-2 is closest to: (   )
(**a**) 0 V;   (**b**) 10 mV;   (**c**) 30 mV;   (**d**) 58 mV.

6. Explain how you worked out the experimental value of r_{out} in Table 28-2.

Troubleshooting (Optional)

7. Explain the dc and ac output voltage you got when R_2 was shorted.

8. If there is a collector-emitter short in Fig. 28-1, what happens to the input impedance of the emitter-follower?

Design (Optional)

9. How did you arrive at your design value of R_E?

10. Optional. Instructor's question.

122

EXPERIMENT 29

CLASS B PUSH-PULL AMPLIFIERS

▼

In a class B push-pull amplifier, each transistor operates in the active region for half of the ac cycle. With a single power supply, the ac output compliance is approximately equal to V_{CC}. Class B push-pull amplifiers are widely used for the output stage of audio systems because they can deliver more load power than class A amplifiers. In fact, the theoretical efficiency of a class B push-pull amplifier approaches 78.5 percent, far greater than class A.

The main problem with class B push-pull amplifiers is setting up a stable Q point near cutoff. The required V_{BE} drop for each transistor is in the vicinity of 0.6 to 0.7 V, with the exact value determined by the quiescent collector current needed to avoid crossover distortion. Since collector current increases by a factor of 10 for each V_{BE} increase of 60 mV, setting up a stable and precise I_{CQ} is difficult. Voltage-divider bias is not practical because the Q point is too sensitive to changes in supply voltage, temperature, and transistor replacement. As discussed in the textbook, there is a real danger of thermal runaway. Diode bias is the usual way to bias a class B push-pull amplifier. The idea is to use diodes whose IV curves match the IV curves of the emitter diodes.

In this experiment you will build a class B push-pull amplifier with voltage-divider bias and with diode bias. This will allow you to see that diode bias is more stable than voltage-divider bias.

REQUIRED READING

Chapter 12 (Sec. 12-7 to 12-10) of *Electronic Principles*, 5th ed.

EQUIPMENT

1 audio generator
1 power supply: adjustable from 0 to 15 V
2 diodes: 1N914
2 transistors: 2N3904, 2N3906
5 ½-W resistors: 100 Ω, two 680 Ω, two 4.7 kΩ
3 capacitors: two 1 μF, 100-μF (16-V rating or better)
2 VOMs (or a dc milliammeter and dc voltmeter)
1 oscilloscope

PROCEDURE

Crossover Distortion

1. Adjust your power supply to 5 V and then connect the circuit of Fig. 29-1*a* with this supply voltage.
2. Set the input frequency to 1 kHz and the signal level across the audio generator at 2 V p-p.
3. Look at the output signal across the load resistor (100 Ω). You should see crossover distortion.

Sensitivity of Voltage-Divider Bias

4. Reduce the signal level to zero and connect the VOM as an ammeter in series with the upper collector (see Fig. 29-1*b*).

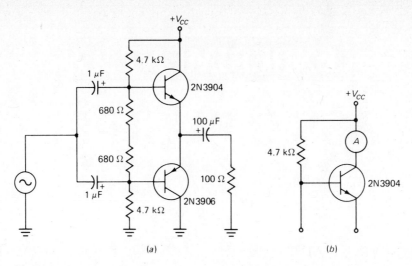

(a)

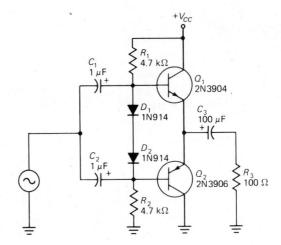

(b)

Figure 29-1

5. Connect a second VOM as a voltmeter across the power supply.
6. Adjust the V_{CC} supply to get an I_{CQ} of 10 μA. Record the value of V_{CC} in Table 29-1.
7. Adjust the V_{CC} supply to set up the other values of I_{CQ}. Record the values of V_{CC}.
8. Reduce V_{CC} to zero. Calculate the sensitivity of collector current to supply voltage with this formula:

$$\text{Sensitivity} = \frac{\max I_{CQ} - \min I_{CQ}}{\max V_{CC} - \min V_{CC}} \quad (29\text{-}1)$$

Use the maximum and minimum values in Table 29-1. Record your answer in Table 29-1.

MPP Value

9. Adjust V_{CC} to get approximately 1 mA of quiescent collector current. Remove the ammeter from the upper collector circuit and reconnect the collector to the supply voltage.
10. Increase the audio generator until the output signal just starts clipping. Back off slightly from this level so that the output peaks are no longer clipped. If your circuit is working correctly, the peak-to-peak output should be approximately equal to V_{CC} (around 10 V). This is the MPP value of the circuit.

Sensitivity of Diode Bias

11. Connect the circuit of Fig. 29-2.
12. Reduce the audio generator to zero. Measure and record V_{CC} for an I_{CQ} of 10 μA, 100 μA, and 1 mA. If you cannot set up a particular I_{CQ} value, ignore the V_{CC} reading. (Use Table 29-2.)
13. Set V_{CC} to 15 V. Record I_{CQ} in Table 29-2. Remove the ammeter and reconnect the collector to the supply voltage.

Figure 29-2

14. Calculate the sensitivity of collector current to supply voltage using Eq. (29-1). Record your answer in Table 29-2.

Troubleshooting (Optional)

15. Ask the instructor to insert a trouble in the circuit of Fig. 29-2.
16. Locate and repair the trouble.
17. Repeat Steps 15 and 16 as often as indicated by the instructor.

Computer (Optional)

18. Write and run a program that prints out $P_{L(\max)}$, P_S, and η for the circuit of Fig. 29-2. The inputs are V_{CC}, R_1, R_2, and R_L.

DATA FOR EXPERIMENT 29

Table 29-1. Sensitivity of Voltage-Divider Bias

I_{CQ}	V_{CC}
10 μA	
100 μA	
1 mA	
10 mA	
Sensitivity =	

Table 29-2. Sensitivity of Diode Bias

I_{CQ}	V_{CC}
10 μA	
100 μA	
1 mA	
	15 V
Sensitivity =	

QUESTIONS FOR EXPERIMENT 29

1. The measured V_{CEQ} of each transistor in Table 29-1 was closest to: ( )
   (a) 5 V; (b) 10 V; (c) 15 V; (d) 20 V.
2. The MPP value in Table 29-2 is closest to: ( )
   (a) 5 V; (b) 10 V; (c) 15 V; (d) 20 V.
3. The theoretical current drain of Fig. 29-2 is approximately: ( )
   (a) 0.915 mA; (b) 15.9 mA; (c) 16.8 mA; (d) 20 mA.
4. The class B push-pull amplifier of Fig. 29-2 has a theoretical efficiency of ( )
   approximately:
   (a) 1%; (b) 25%; (c) 75%; (d) 100%.
5. Briefly explain what the calculated sensitivities of Tables 29-1 and 29-2 mean.

6. Explain how the class B push-pull amplifier of Fig. 29-2 works.

Troubleshooting (Optional)

7. Someone has built the circuit of Fig. 29-1 using a supply voltage of $+15$ V. Then he or she discovers one of the transistors has been destroyed. Explain what has happened.

8. Suppose a collector-emitter short occurs in Q_1 of Fig. 29-2. Explain what will happen to Q_2.

9. Optional. Instructor's question.

10. Optional. Instructor's question.

EXPERIMENT 30

THE ZENER FOLLOWER

▼

By cascading a zener diode and an emitter follower, we can build voltage regulators with large load currents. An improved regulator like this can hold the load voltage almost constant despite large changes in load current because the circuit appears stiff over a greater range of load resistance. The zener follower is an example of a series regulator, one whose load current flows through a pass transistor. Because of their simplicity, series regulators are widely used.

In this experiment you will build a regulated power supply consisting of a bridge rectifier, a capacitor-input filter, and a zener follower.

REQUIRED READING

Chapter 12 (Sec. 12-9) of *Electronic Principles*, 5th ed.

EQUIPMENT

1 center-tapped transformer, 12.6 V ac (Triad F-25X or equivalent) with fused line cord
4 silicon diodes: 1N4001 (or equivalent)
1 zener diode: 1N757 (or equivalent 9-V zener diode)
1 transistor: 2N3055 (or equivalent power transistor)
2 ½-W resistors: 1 kΩ, 10 kΩ
2 1-W resistors: 100 Ω, 470 Ω
1 capacitor: 470 μF (25-V rating or better)
1 VOM (analog or digital multimeter)
1 oscilloscope

PROCEDURE

Regulated Power Supply

1. A 1N757 has a nominal zener voltage of 9.1 V. In Fig. 30-1, calculate the input voltage, zener voltage, and output voltage for the zener follower. (The input voltage is across the filter capacitor.) Record your answers in Table 30-1.
2. Connect the regulated power supply of Fig. 30-1.
3. Measure and record all voltages listed in Table 30-1.

Voltage Regulation

4. Estimate and record the output voltages in Fig. 30-1 for each of the load resistors listed in Table 30-2.
5. Measure and record the output voltages for the load resistances of Table 30-2.

Ripple Attenuation

6. For each load resistance listed in Table 30-3, calculate and record the peak-to-peak ripple across the filter capacitor. Also calculate and record the peak-to-peak ripple at the output. (Assume a zener resistance of 10 Ω.)
7. For each load resistance of Table 30-3, measure and record the peak-to-peak ripple at the input and output of the zener follower. (Note: If the ripple appears very fuzzy or erratic, you may have parasitic oscillations, an undesirable phenomenon that is discussed in Sec. 22-7 of Chap. 22. Try shortening the leads. If that doesn't work, consult the instructor.)

Troubleshooting (Optional)

8. Assume D_1 is open.
9. Estimate the input and output voltage of the zener follower for the foregoing trouble. Record your answers in Table 30-4.

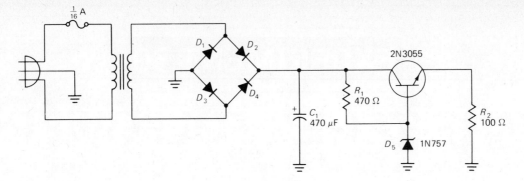

Figure 30-1

10. Connect the circuit with the foregoing trouble. Measure and record the input and output voltage.
11. Repeat Steps 9 and 10 for the other troubles listed in Table 30-4.

Design (Optional)

12. Modify the power supply so that it produces a regulated output voltage of approximately 5.5 V for a load current between 0 and 55 mA. The zener current should be approximately 20 mA. Record your design changes in Table 30-5.
13. Calculate and record the input voltage, input ripple, output voltage, and output ripple for the zener follower with a load resistance of 100 Ω. Assume the zener diode has a zener resistance of 7 Ω.
14. Connect your design with a load resistance of 100 Ω. Measure all dc voltages and ripples listed in Table 30-5. Record your data.

Computer (Optional)

15. Debug the following program. Then enter and run it.

```
10 PRINT "ENTER ZENER VOLTAGE"
20 INPUT VZ
30 IS = (17.8 + VZ)/6800: IL = VZ/RL: VOUT
   = VZ + 0.7
40 PRINT "IS =": PRINT IS
50 PRINT "IL = ": PRINT IL
60 PRINT "VOUT =": PRINT VOUT
70 STOP
```

16. Write and run a program that prints out the power dissipation of the pass transistor. The inputs are V_{in}, V_{out}, and R_L.

DATA FOR EXPERIMENT 30

Table 30-1. Regulated Power Supply

	V_{in}	V_Z	V_{out}
Calculated			
Measured			

Table 30-2. Voltage Regulation

	Estimated V_{out}	Measured V_{out}
R_L		
100 Ω		
1 kΩ		
10 kΩ		

Table 30-3. Ripple

	Calculated V_{rip}		Measured V_{rip}	
R_L	In	Out	In	Out
100 Ω				
1 kΩ				
10 kΩ				

Table 30-4. Troubleshooting

	Estimated		Measured	
	V_{in}	V_{out}	V_{in}	V_{out}
Open D_1				
Open R_1				
Shorted D_5				
Collector-emitter open				

Table 30-5. Design

Describe your changes here:

	V_{in}	Input V_{rip}	V_{out}	Output V_{rip}
Calculated				
Measured				

QUESTIONS FOR EXPERIMENT 30

1. When the load resistance increases in Table 30-2, the measured output voltage: ( )
   (a) decreases slightly: (b) stays the same; (c) increases slightly;
   (d) none of the foregoing.

2. When the load resistance increases, the input ripple to the zener follower: ( )
   (a) decreases; (b) stays the same; (c) increases; (d) none of the foregoing.

3. The pass transistor of Fig. 30-1 has a power dissipation that is closest to: ( )
   (a) 0.25 W; (b) 0.5 W; (c) 0.7 W; (d) 1 W.

4. The zener diode of Fig. 30-1 has a zener current of approximately: ( )
   (a) 1 mA; (b) 2.35 mA; (c) 12.2 mA; (d) 18.5 mA.

5. In Fig. 30-1, the load current is approximately; ( )
   (a) 1 mA; (b) 18.5 mA; (c) 84 mA; (d) 523 mA.

6. Explain how the regulated power supply of Fig. 30-1 works.

Troubleshooting (Optional)

7. Explain why the regulator continues to work even though D_1 is opened.

8. Suppose the circuit of Fig. 30-1 has a collector-emitter short (base, emitter, and collector shorted together). What components are likely to be destroyed?

Design (Optional)

9. What changes did you make and why did you make them?

10. Optional. Instructor's question.

AN AUDIO AMPLIFIER

▼

In this experiment you will build a discrete audio amplifier with a class A input stage and a class B push-pull emitter follower as shown in Fig. 31-1. Adjustment R_7 is included to allow you to set the Q_3 emitter voltage at +5 V, half the supply voltage. This lets the output signal swing equally in both directions to get maximum MPP value. Some of the resistance values have not been optimized to allow you to improve the design (optional). The circuit is fairly complicated, so take your time in wiring it. Double-check your connections before applying power.

▲   ▲

REQUIRED READING

Chapters 10 to 12 of *Electronic Principles*, 5th ed.

EQUIPMENT

1 audio generator
1 power supply: 10 V
2 diodes: 1N914
4 transistors: three 2N3904, 2N3906
9 ½-W resistors: two 100 Ω, two 1 kΩ, 2.2 kΩ, 3.6 kΩ, 4.7 kΩ, two 10 kΩ
1 potentiometer: 5 kΩ
4 capacitors: two 1 µF, 100 µF, 470 µF (16-V rating or better)

1 VOM (analog or digital multimeter)
1 oscilloscope

PROCEDURE

Audio Amplifier

1. Assume the base resistance of Q_2 is adjusted to produce a quiescent voltage of +5 V at the emitter of Q_3. Calculate and record all dc voltages listed in Table 31-1.
2. Assume an ac load voltage of 4 V p-p. Also assume all h_{fe} values are 100. Calculate and record all ac voltages listed in Table 31-2.
3. Build the audio amplifier of Fig. 31-1. Adjust the

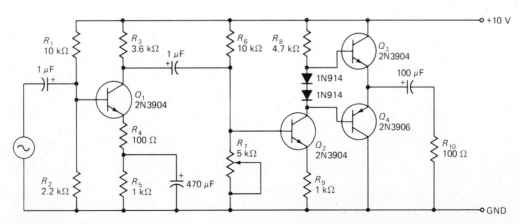

Figure 31-1

base resistor of Q_2 to produce a dc voltage $+5$ V at the emitter of Q_3.

4. Set the input frequency to 1 kHz and adjust the ac load voltage to 4 V p-p.
5. Use the VOM to measure all dc voltages listed in Table 31-1. Use the oscilloscope to measure all ac voltages in Table 31-2. Record your data.

Troubleshooting (Optional)

6. Ask the instructor to insert a trouble in your circuit.
7. Locate and repair the trouble.
8. Repeat Steps 6 and 7 as often as indicated by the instructor.

Design (Optional)

9. Measure the MPP value. Notice that it is much less than 10 V, the supply voltage. Record the initial MPP in Table 31-3.

10. Try to figure out how to increase the MPP value. For instance, changing certain resistances will increase the MPP value.
11. Insert your changes and measure the MPP value.
12. When you have increased the MPP value as much as possible, record the final MPP in Table 31-3. Also record the changes you inserted.

Computer (Optional)

13. Write and run a program that calculates all dc voltages in Fig. 31-1. Assume the variable resistor of Q_2 has been adjusted to produce $+5$ V at the Q_3 emitter.

● **DATA FOR EXPERIMENT 31**

Table 31-1. DC Voltages

	Calculated			Measured		
	B	E	C	B	E	C
Q_1						
Q_2						
Q_3						
Q_4						

Table 31-2. AC Voltages

	Calculated			Measured		
	B	E	C	B	E	C
Q_1						
Q_2						
Q_3						
Q_4						

Table 31-3. Design

Initial MPP = _____

Final MPP = _____

Changes were as follows:

_____

QUESTIONS FOR EXPERIMENT 31

1. Resistor R_7 is adjusted to get approximately: ( )
   (a) 10 mA through R_8; (b) +10 V at the collector of Q_3; (c) +5 V at the emitter of Q_3; (d) 0 V at the collector of Q_4.
2. The capacitive reactance of 100 μF at 1 kHz is approximately: ( )
   (a) 1.59 Ω; (b) 6.28 Ω; (c) 100 Ω; (d) 1 kΩ.
3. The amplifier of Fig. 31-1 had a voltage gain closest to: ( )
   (a) 1; (b) 25; (c) 100; (d) 200.
4. The measured ac voltage at the base of Q_3 was slightly higher than the ac ( )
   output voltage because:
   (a) the output capacitor was too small; (b) of the drop across r'_e; (c) R_7 was adjustable; (d) Q_2 was an *npn* transistor.

5. One reason the MPP value of the output stage is less than V_{CC} is because of: (   )
   (a) voltage drop across X_C;   (b) offset voltage of the 1N914s;   (c) V_{BE} drops of the output transistors;   (d) power dissipation by the load resistor.
6. Explain how the audio amplifier of Fig. 31-1 works.

Troubleshooting (Optional)

7. Suppose a 1N914 shorts in Fig. 31-1. What kind of symptoms will this produce?

8. Resistor R_4 of Fig. 31-1 is shorted by a solder bridge. Describe some of the symptoms that result.

Design (Optional)

9. Why is the MPP value less than V_{CC} in Fig. 31-1?

10. Optional. Instructor's question.

EXPERIMENT 32

JFET CURVES

▼

The junction field-effect transistor (JFET) has drain curves that resemble the collector curves of a bipolar transistor. The JFET also has a transconductance curve that is the graph of drain current versus gate-source voltage. In this experiment, you will build a circuit that displays JFET drain and transconductance curves.

REQUIRED READING

Chapter 13 (Secs. 13-1 to 13-4) of *Electronic Principles*, 5th ed.

EQUIPMENT

1 audio generator
1 power supply: adjustable from at least 1 to 10 V
1 JFET: MPF102 (or any *n*-channel JFET with an I_{DSS} greater than 2 mA)

1 diode: 1N914 (or any small-signal silicon diode)
2 ½-W resistors: 100 Ω, 1 MΩ
1 capacitor: 1 µF
1 oscilloscope

PROCEDURE

1. Set up the oscilloscope as follows. Set the horizontal sensitivity to 1 V/cm (dc input) and the vertical sensitivity to 0.1 V/cm (dc input). Center the spot in the upper left-hand corner.

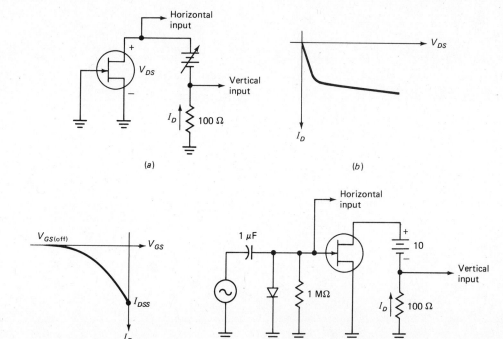

Figure 32-1

2. Connect the circuit of Fig. 32-1*a*.
3. Slowly vary the supply from minimum up to 10 V. If the spot deflects off the screen, change the vertical sensitivity as needed to keep it on.
4. Vary the supply voltage back and forth rapidly from minimum to 10 V. As you do this, notice you get a drain curve similar to Fig. 32-1*b*.
5. Estimate the pinchoff voltage. Write the value here: $V_P = $ _____.
6. Adjust the supply to get a V_{DS} greater than pinchoff. Then, figure out the relation between I_{DSS} and vertical deflection. Write the value of I_{DSS} here: _____.
7. Disconnect the oscilloscope. Center the spot in the upper right-hand corner.

8. Connect the circuit of Fig. 32-1*d*.
9. Set the audio generator to 100 Hz. Increase the signal from zero upward to get a transconductance curve like Fig. 32-1*c*. (You may have to readjust the scope sensitivities.)
10. Estimate $V_{GS(\text{off})}$ and record here: _____.
11. Change the horizontal and vertical sensitivity as needed to get as large a transconductance curve as possible on the screen. Notice how the curve has a parabolic shape.
12. If a curve tracer is available, look at the drain curves of the MPF102.

QUESTIONS FOR EXPERIMENT 32

1. In Fig. 32-1a, each milliampere of drain current produces a vertical input of: (   )
   (a) 1 mV; (b) 10 mV; (c) 100 mV; (d) 1 V.

2. If the I_{DSS} of the JFET is 16 mA, the vertical input in Fig. 32-1a for a V_{DS} (   )
   greater than pinchoff will be:
   (a) 16 μV; (b) 16 mV; (c) 1.6 V; (d) 16 V.

3. The drain curve appeared upside-down in this experiment because: (   )
   (a) an n-channel JFET was used; (b) the voltage across the drain-source
   terminals is negative; (c) the negative end of V_{DD} is connected to the
   vertical input; (d) the voltage across the drain-source terminals is positive.

4. The pinchoff voltage was in the vicinity of: (   )
   (a) 0 to 1 V; (b) 2 to 8 V; (c) 8 to 12 V; (d) 12 to 16 V.

5. The magnitude of pinchoff voltage theoretically equals the magnitude of: (   )
   (a) 2 V; (b) $V_{GS(off)}$; (c) $V_{GS(off)}/4$; (d) V_{DS}.

6. In Fig. 32-1d, the input audio signal is: (   )
   (a) negatively clamped; (b) positively clamped; (c) negatively clipped;
   (d) negatively peak-rectified.

7. The voltage applied to the horizontal input in Fig. 32-1d is: (   )
   (a) V_{GS}; (b) V_{DS}; (c) $100I_D$; (d) none of these.

8. If the horizontal input is disconnected in Fig. 32-1d and the audio generator (   )
   is turned down to zero, you will see a;
   (a) vertical line; (b) horizontal line; (c) spot at zero; (d) vertically deflected spot.

9. Optional. Instructor's question.

10. Optional. Instructor's question.

EXPERIMENT 33

JFET BIAS

▼

Gate bias is the simplest but worst way to bias a JFET for linear operation because the drain current depends on the exact value of V_{GS}. Since V_{GS} has a large variation, the drain current has a large variation. Self-bias offers some improvement because the source resistor produces local feedback which reduces the effect of V_{GS}. When large supply voltages are available, voltage-divider bias results in a relatively stable Q point. Finally, current-source bias can produce the most stable Q point because a bipolar current source sets up the drain current through the JFET.

In this experiment, you will bias a JFET in the different methods just described. This will illustrate the stability of each type of bias.

REQUIRED READING

Chapter 14 (Secs. 14-1, 14-2, 14-3, and 14-9) of *Electronic Principles*, 5th ed.

EQUIPMENT

2 power supplies: adjustable from 0 to ± 15 V
1 transistor: 2N3904
3 JFETs: MPF102 (or equivalent)
7 ½-W resistors: 470 Ω, 680 Ω, 1 kΩ, 2.2 kΩ, 6.8 kΩ, 33 kΩ, 100 kΩ
1 VOM (analog or digital multimeter)

PROCEDURE

Measuring I_{DSS}

1. Refer to the data sheet of an MPF102 in the Appendix. Notice that $V_{GS(\text{off})}$ has a maximum of -8 V; the minimum value is not specified. Also notice that I_{DSS} has a minimum value of 2 mA and a maximum value of 20 mA.
2. Connect the circuit of Fig. 33-1a. Measure the drain current. Record this value of I_{DSS} in Table 33-1. (Note: Because of heating effects, the drain current

may decrease slowly. Take your reading as soon after power-up as possible.)
3. Repeat Step 2 for the other JFETs.

Gate Bias

4. With gate bias, you apply a fixed gate voltage that reverse-biases the gate of the JFET. This produces a drain current that is less than I_{DSS}. The problem is that you cannot accurately predict the drain current in mass production because of the variation in the required V_{GS}. The following steps will illustrate this point.

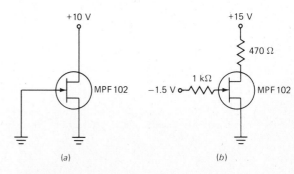

Figure 33-1

5. Connect the circuit of Fig. 33-1b. Measure V_{GS}, I_D, and V_D. Record the data in Table 33-2.
6. Repeat Step 5 for the other JFETs. If you have a random set of three JFETs, the drain current usually will show a significant variation from one JFET to another.

Measuring $V_{GS(off)}$

7. Here is an approximate way to measure $V_{GS(off)}$. Insert the first JFET into the gate-biased circuit. Increase the negative gate supply voltage of Fig. 33-1b until the drain current is approximately 1 μA. (If your VOM cannot measure down to 1 μA, then use 10 μA or 100 μA, or whatever low value your instructor indicates.) Record the approximate $V_{GS(off)}$ in Table 33-1.
8. Repeat Step 7 for the other two JFETs.

Self-Bias

9. The data sheet of a JFET lists a maximum I_{DSS} of 20 mA and a maximum $V_{GS(off)}$ of −8 V. If a JFET has these values, its transconductance curve appears as shown in Fig. 33-2a. Notice that the approximate source resistance for self-bias is

$$R_S = \frac{8 \text{ V}}{20 \text{ mA}} = 400 \text{ }\Omega$$

The minimum I_{DSS} is 2 mA. For this experiment, we will assume the minimum $V_{GS(off)}$ is −2 V. A JFET with these values has the lower transconductance curve of Fig. 33-2a, and required R_S is 1 kΩ. An average source resistance is around 700 Ω, so we will use 680 Ω in our test circuit.
10. Assume a V_{GS} of −2 V in Fig. 33-2b. Calculate I_D, V_D, and V_S. Record your answers in Table 33-3.
11. Connect the self-bias circuit of Fig. 33-2b. Measure I_D, V_D, and V_S. Record your data in Table 33-3.
12. Repeat Step 11 for the other JFETs.

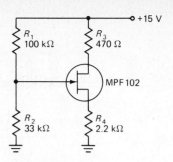

Figure 33-3

13. Notice that the drain current of the self-bias circuit (Table 33-3) has less variation than the drain current of the gate-biased circuit (Table 33-2).

Voltage-Divider Bias

14. Assume V_{GS} is a −2 V in Fig. 33-3. Calculate I_D, V_D, and V_S. Record your answers in Table 33-4.
15. Connect the circuit. Measure and record, the quantities of Table 33-4.
16. Repeat Step 15 for the other JFETs. Notice that the drain current of Table 33-4 has less variation than the drain currents of Tables 33-2 and 33-3.

Current-Source Bias

17. Assume V_{GS} is −2 V in Fig. 33-4. Calculate and record the quantities of Table 33-5.
18. Connect the circuit. Measure I_D, V_D, and V_S. Record your data.

Troubleshooting (Optional)

19. Assume R_2 is shorted in Fig. 33-4. Calculate and record V_D in Table 33-6.
20. Insert the trouble into your circuit. Measure and record V_D.

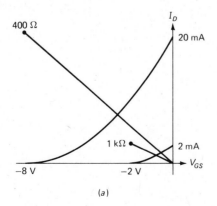

(a)

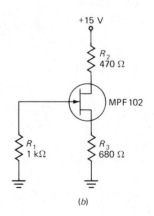

(b)

Figure 33-2

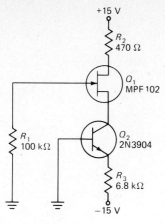

+15 V

R_2 470 Ω

Q_1 MPF 102

R_1 100 kΩ

Q_2 2N3904

R_3 6.8 kΩ

−15 V

Figure 33-4

21. Repeat Steps 19 and 20 for the other troubles listed in Table 33-6.

Design (Optional)

22. Use Eq. (14-2) of your textbook and the data of Table 33-1 to calculate the source resistance of a self-biased circuit. Average the three source resistances. Record your answer at the top of Table 33-7.

23. Connect the circuit of Fig. 33-2b with your design value of R_S. Measure I_D, V_D, and V_S. Record data.

24. Repeat Step 23 for the other JFETs.

Computer (Optional)

25. Enter and run this program:

```
10 FOR X = 1 TO 10
20 PRINT X
30 NEXT X
40 END
```

26. A JFET has an I_{DSS} of 20 mA and a $V_{GS(\text{off})}$ of −8 V. Write and run a program that prints out I_D for these values of V_{GS}: 0, −1 V, −2 V, −3 V, . . ., −8 V.

DATA FOR EXPERIMENT 33

Table 33-1. JFET Data

JFET	I_{DSS}	$V_{GS(off)}$
1		
2		
3		

Table 33-2. Gate Bias: $V_{GG} = -1.5$ V

JFET	V_{GS}	I_D	V_D
1			
2			
3			

Table 33-3. Self-Bias

JFET	Calculated			Measured		
	I_D	V_D	V_S	I_D	V_D	V_S
1						
2						
3						

Table 33-4. Voltage-Divider Bias

JFET	Calculated			Measured		
	I_D	V_D	V_S	I_D	V_D	V_S
1						
2						
3						

Table 33-5. Current-Source Bias

JFET	Calculated			Measured		
	I_D	V_D	V_S	I_D	V_D	V_S
1						
2						
3						

Table 33-6. Troubleshooting

	Estimated		Measured	
Trouble	V_D	V_C	V_D	V_C
R_2 shorted				
Q_2 collector-emitter short				
R_3 open				

Table 33-7. Design: R_S = _____

JFET	I_S	V_D	V_S
1			
2			
3			

QUESTIONS FOR EXPERIMENT 33

1. This experiment proved that the circuit with the least variation in drain ( )
   current was:
   (a) gate bias; (b) self-bias; (c) voltage-feedback; (d) current-source
   bias.
2. With gate bias, the drain current has which of the following: ( )
   (a) almost constant value; (b) constant drain voltage; (c) values greater
   than I_{DSS}; (d) large variations.
3. Self-bias is better than: ( )
   (a) gate bias; (b) voltage-divider bias; (c) current-source bias;
   (d) emitter bias
4. The voltage across the source resistor of Fig. 33-3 equals the gate voltage plus ( )
   the magnitude of:
   (a) V_{GS}; (b) V_D; (c) V_S; (d) V_{DD}.
5. The voltage-divider bias of Fig. 33-3 would be more stable if we: ( )
   (a) decreased the supply voltage; (b) increased the supply voltage;
   (c) decreased R_2; (d) increased R_1.
6. Briefly discuss the variations in drain current for the four types of JFET bias.

Troubleshooting (Optional)

7. Suppose the voltage across the source resistor of Fig. 33-3 is zero. Name three possible
   troubles.

8. The drain voltage of Fig. 33-2 is 15 V. Name three possible troubles.

Design (Optional)

9. What value of R_S did you use in your design? How did you arrive at this value?

10. Optional. Instructor's question.

JFET AMPLIFIERS

▼

Because the transconductance curve of a JFET is parabolic, large-signal operation of a CS amplifier produces square-law distortion. This is why a CS amplifier is usually operated small-signal. JFET amplifiers cannot compete with bipolar amplifiers when it comes to voltage gain. Because g_m is relatively low, the typical CS amplifier has a relatively low voltage gain.

The CD amplifier, better known as the source follower, is analogous to the emitter follower. The voltage gain approaches unity and the input impedance approaches infinity, limited only by the external biasing resistors connected to the gate. The source follower is a popular circuit often found near the front end of measuring instruments.

In this experiment, you will build a CS amplifier and a source follower to verify the relations discussed in your textbook.

REQUIRED READING

Chapter 14 (Sec. 14-5) of *Electronic Principles*, 5th ed.

EQUIPMENT

1 audio generator
1 power supply: 15 V
3 JFETs: MPF102 (or equivalent)
4 ½-W resistors: 1 kΩ, two 2.2 kΩ, 220 kΩ
3 capacitors: two 1 µF, 100 µF (16-V rating or better)
1 potentiometer: 5 kΩ
1 oscilloscope

PROCEDURE

CS Amplifier

1. Assume the JFET of Fig. 34-1 has a typical g_m of 2000 µS. Calculate the unloaded voltage gain, output voltage, and output impedance. Record your answers in Table 34-1.
2. Connect the circuit with R_L equal to infinity (no load resistor).
3. Adjust the audio generator to 1 kHz. Set the signal level to 0.1 V p-p across the input.

4. Look at the output signal. It should be an amplified sine wave. Measure and record the peak-to-peak output voltage. Then calculate the voltage gain. Record your answer as the measured A in Table 34-1.
5. Connect the 5-kΩ potentiometer as a variable load resistance. Adjust this load resistance until the output voltage is half of the unloaded output voltage.
6. Disconnect the 5-kΩ potentiometer and measure its resistance. Record this as r_{out} in Table 34-1. (Note: You have just found the Thevenin or output impedance by the matched-load method.)
7. Repeat Steps 1 to 6 for the other JFETs.

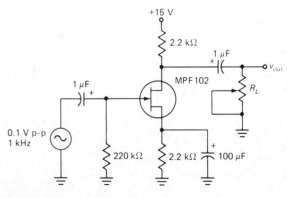

Figure 34-1

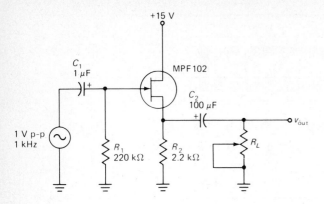

Figure 34-2

Source Follower

8. Assume a typical g_m of 2000 μS in Fig. 34-2. Calculate the unloaded voltage gain, output voltage, and output impedance. Record your answers in Table 34-2.
9. Connect the circuit with R_L equal to infinity. Adjust the frequency to 1 kHz and the signal level to 1 V p-p across the input.
10. Measure and record the output voltage. Calculate the voltage gain and record as the measured A in Table 34-2.
11. Measure and record the output impedance by the matched-load method used earlier.
12. Repeat Steps 8 to 11 for the other JFETs.

Troubleshooting (Optional)

13. Table 34-3 lists dc and ac symptoms for Fig. 34-2. Try to figure out what trouble would produce these symptoms. When you think you have the answer, insert the trouble into the circuit. Then measure the dc and ac voltages to verify that your trouble is causing the symptoms. Record the trouble in Table 34-3.
14. Repeat Step 13 for the other symptoms listed in Table 34-3.

Design (Optional)

15. Redesign the source follower of Fig. 34-2 so that it uses voltage-divider bias. Assume V_{GS} is -2 V and select R_1 and R_2 to produce a V_S of $+7.5$ V. Record your resistance values at the top of Table 34-4.
16. Connect your redesigned source follower. Measure the dc voltage at the source. Record V_S in Table 34-4. Also, set the ac input voltage to 1 V p-p. Measure the ac output voltage. Calculate and record the unloaded voltage gain.
17. Repeat Step 16 for each JFET.

Computer (Optional)

18. Write and run a program that calculates the unloaded voltage gain and output impedance for a source follower like Fig. 34-2. The inputs are g_m and R_S.

Table 34-1. CS Amplifier

JFET	v_{out}	Calculated A	r_{out}	v_{out}	Measured A	r_{out}
1						
2						
3						

Table 34-2. Source Follower

JFET	v_{out}	Calculated A	r_{out}	v_{out}	Measured A	r_{out}
1						
2						
3						

Table 34-3. Troubleshooting

V_G	DC Symptoms V_D	V_S	v_g	v_d	AC Symptoms v_s	v_{out}	Trouble
0	15 V	3.7 V	1 V	0	0	0	
0	15 V	3.7 V	1 V	0	0.82 V	0	
0	15 V	3.7 V	0	0	0	0	
0	0	0	1 V	0	0	0	

Table 34-4. Design. $R_1 =$ _____; $R_2 =$ _____

JFET	V_S	A
1		
2		
3		

QUESTIONS FOR EXPERIMENT 34

1. The calculated voltage gain of Table 34-1 is approximately: ( )
   (a) 0.44; (b) 1; (c) 4.4; (d) 9.4.
2. The output impedance of Fig. 34-1 is closest to: ( )
   (a) 407 Ω; (b) 2.2 kΩ; (c) 5 kΩ; (d) 220 kΩ.
3. The voltage gain of the source follower was closest to: ( )
   (a) 0.5; (b) 0.8; (c) 1; (d) 4.4.

4. The source follower had an output impedance closest to: ( )
   (a) 0; (b) 100 Ω; (c) 200 Ω; (d) 400 Ω.

5. The main advantage of a JFET amplifier is its: ( )
   (a) high voltage gain; (b) low drain current; (c) high input impedance;
   (d) high transconductance.

6. Compare the voltage gain of a CS amplifier like Fig. 34-1 to a bipolar CE amplifier.

Troubleshooting (Optional)

7. In Fig. 34-2, all dc voltages are normal. The ac gate voltage and source voltage are normal. There is no output voltage. What is the most likely trouble?

8. All dc voltages are normal in Fig. 34-2. All ac voltages are zero. What is the most likely trouble?

Design (Optional)

9. How did you arrive at your values of R_1 and R_2?

10. Optional. Instructor's question.

EXPERIMENT 35

JFET APPLICATIONS

▼

One of the main applications of JFETs is the analog switch. In this application, a JFET acts either like an open switch or like a closed switch. This allows us to build circuits that either transmit an ac signal or block it from the output terminals.

In the ohmic region, a JFET acts like a voltage-variable resistance instead of a current source. This means we can change the value of $r_{ds(on)}$ by changing V_{GS}. When a JFET is used as a voltage-variable resistance, the ac signal should be small, typically less than 100 mV.

Another application of the JFET is with automatic gain control (AGC). Because the g_m of a JFET varies with the Q point, we can build amplifiers whose voltage gain is controlled by an AGC voltage.

In this experiment, you will build various JFET circuits to see how the JFET can act as a switch, voltage-variable resistance, and AGC device.

REQUIRED READING

Chapter 14 (Secs. 14-6 and 14-11) of *Electronic Principles*, 5th ed.

EQUIPMENT

1 audio generator
2 power supplies: adjustable to ± 15 V
1 diode: 1N914
3 JFETs: MPF102 (or equivalent)
4 ½-W resistors: 2.2 kΩ, 10 kΩ, two 100 kΩ
2 capacitors: 1 μF
1 switch: SPST
1 VOM (analog or digital multimeter)
1 oscilloscope

PROCEDURE

Analog Switch

1. Measure the approximate $r_{DS(on)}$ of each JFET as follows. Short the gate and source together. Connect the positive lead of an ohmmeter to the drain, and the negative lead to the source. Record the values of $r_{DS(on)}$ in Table 35-1. Throughout this experi-

ment, you may use $r_{DS(on)}$ as an approximation for $r_{ds(on)}$.

2. In Fig. 35-1, calculate v_{out} for each JFET when V_{GS} is zero. Record your answers in Table 35-1.
3. Connect the circuit with the ac signal source at 100 mV p-p and 1 kHz.
4. Measure and record the ac output voltage with S_1 open and S_1 closed.
5. Repeat Step 4 for the other two JFETs.

JFET Chopper

6. What do you think the output voltage of Fig. 35-2 will look like? Sketch the expected waveform in Table 35-2.
7. Connect the circuit with the specified ac input voltages and frequencies.
8. Look at the output voltage. Set the trigger select of the oscilloscope to the 1-kHz signal. Slowly vary the 1-kHz frequency until you see a steady chopped waveform. Sketch the waveform in Table 35-2.

Voltage-Variable Resistance

9. As V_{GG} is varied from zero to a value more negative than $V_{GS(off)}$, the peak-to-peak output voltage of Fig.

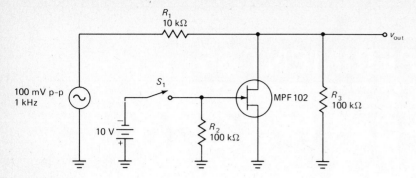

Figure 35-1

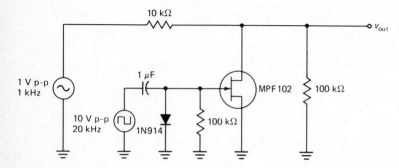

Figure 35-2

35-3 will change. Do you think it will increase or decrease?

10. Connect the circuit.

11. Set up each value of V_{GG} shown in Table 35-3. Measure and record the ac output voltage. Calculate and record the value of $r_{ds(on)}$.

AGC Curcuit

12. Connect the circuit of Fig. 35-4. Adjust V_{GG} to get maximum output signal. Measure and record v_{out} and V_{GG} in Table 35-4.

13. Adjust V_{GG} until v_{out} drops in half. Measure and record v_{out} and V_{GG}.

14. Repeat Step 13 two times.

Troubleshooting (Optional)

15. Assume V_{GG} is set to produce maximum output in Fig. 35-4. For each set of dc and ac symptoms listed in Table 35-5, figure out what the corresponding trouble may be. Insert your suspected trouble, then check the dc and ac voltages. When you locate each trouble, record it in Table 35-5.

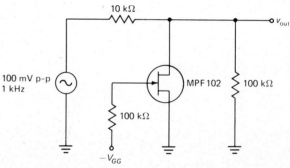

Figure 35-3

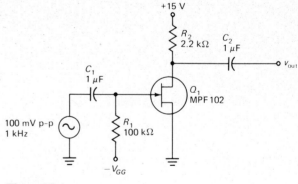

Figure 35-4

Design (Optional)

16. Select a value of R_1 in Fig. 35-1 that increases the attenuation by a factor of 10 when S_1 is open. Connect and test the circuit with your design value.

Computer (Optional)

17. Write and run a program that calculates the values of $r_{ds(on)}$ in Table 35-3. For the inputs, use any data in Fig. 35-3 and Table 35-3.

DATA FOR EXPERIMENT 35

Table 35-1. JFET Analog Switch

JFET	Measured $r_{DS(on)}$	Calculated v_{out}	Measured v_{out}	
			S_1 open	S_1 closed
1				
2				
3				

Table 35-2. JFET Chopper

Expected waveform (sketch below)   **Experimental waveform**

Table 35-3. Voltage-Variable Resistance

V_{GG}	v_{out}	$r_{ds(on)}$
0		
−0.5 V		
−1 V		
−1.5 V		
−2 V		
−2.5 V		
−3 V		
−3.5 V		
−4 V		
−4.5 V		
−5 V		

Table 35-4. AGC Circuit

Condition	v_{out}	V_{GG}
Maximum output		
Max/2		
Max/4		
Max/8		

Table 35-5. Troubleshooting

DC Symptoms		AC Symptoms		Trouble
V_G	V_D	v_g	v_d	
OK	OK	0	0	
OK	0	OK	0	
0	0	0	0	

QUESTIONS FOR EXPERIMENT 35

1. The JFET analog switch of Fig. 35-1 attenuates the signal when:　( )
   (a) it is on the negative half-cycles;　(b) S_1 is open;　(c) S_1 is closed;
   (d) R_1 is shorted.
2. The gate circuit of Fig. 35-2 contains a:　( )
   (a) 1-kHz frequency;　(b) positive dc clamper;　(c) negative dc clamper;
   (d) forward bias.
3. With the voltage-variable resistance of Fig. 35-3, the output signal increases　( )
   when V_{GS}:
   (a) becomes more negative:　(b) stays the same;　(c) becomes more
   positive;　(d) is zero.
4. In the AGC circuit of Fig. 35-4, the output voltage decreases when:　( )
   (a) the ac input signal increases;　(b) V_{GG} goes to zero;　(c) V_{GG} becomes
   more negative;　(d) none of the foregoing.
5. Explain how the JFET analog switch of Fig. 35-1 works.

6. Explain how the voltage-variable resistance circuit of Fig. 35-3 works.

Troubleshooting (Optional)

7. The dc drain voltage is zero in Fig. 35-4. Name three possible troubles.

8. The output voltage of Fig. 35-1 is zero with S_1 open or closed. Name three possible troubles.

Design (Optional)

9. Which do you think is probably better for a JFET analog switch: a low or high $r_{ds(on)}$ when $V_{GS} = 0$? Explain your reasoning.

10. Optional. Instructor's question.

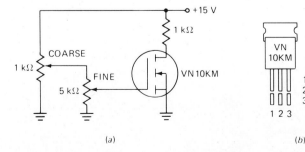

EXPERIMENT 36

VMOS CIRCUITS

▼

The MOSFET has an insulated gate that results in extremely high input impedance. The depletion-type MOSFET, also called a normally on MOSFET, can operate in either the depletion or enhancement mode. The enhancement-type MOSFET, also called a normally off MOSFET, can operate only in the enhancement mode. VMOS transistors are enhancement-type MOSFETs that are useful in applications requiring high load power, including audio amplifiers, RF amplifiers, interfacing, etc. In this experiment you will connect some VMOS circuits to get a better understanding of how they work.

REQUIRED READING

Chapter 13 (Sec. 13-7) and Chapter 14 (Sec. 14-8) of *Electronic Principles*, 5th ed.

EQUIPMENT

1 audio generator
1 power supply: 15 V
1 diode: TIL221 (or similar red LED)
3 MOSFETs: RCA SK9155 or Radio Shack VN10KM (or similar VMOS transistors)
5 ½-W resistors: 330 Ω, 560 Ω, 1 kΩ, two 22 kΩ
2 capacitors: 0.1 μF, 100 μF (16-V rating or better)
1 switch: SPST
2 potentiometers: 1 kΩ, 5 kΩ (or similar low-resistance pots)
1 VOM (analog or digital multimeter)

1 oscilloscope
Graph paper

PROCEDURE

Threshold Voltage

1. Connect the circuit of Fig. 36-1a. (See Fig. 36-1b for VMOS connections.)
2. Most data sheets define threshold voltage as the gate voltage that produces a drain current of 10 μA. Adjust the coarse and fine controls to get a drain current of 10 μA. Record the threshold voltage in Table 36-1.
3. Repeat Steps 1 and 2 for the other VMOS transistors.

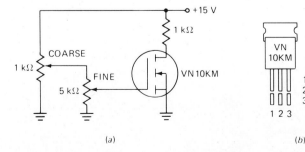

Figure 36-1

Transconductance Curve

4. With any VMOS transistor in the circuit, set the fine control to maximum. Adjust the coarse control to produce a drain current slightly more than 10 mA. Adjust the fine control to get each drain current listed in Table 36-2. Record each gate voltage.

5. Graph your data, I_D versus V_{GS}. Label or otherwise keep track of the VMOS transistor used to get the transconductance curve. You will use this transistor later in the experiment.

Voltage-Divider Bias

6. Assume a V_{GS} of $+2$ V in Fig. 36-2. Calculate I_D, V_G, and V_S. Record your answers in Table 36-3.

7. Connect the circuit. Measure and record I_D, V_G, and V_S.

8. Repeat Steps 6 and 7 for the other VMOS transistors.

Source Follower

9. Calculate the g_m of the transistor in Fig. 36-2. (Use your graph for this.) Calculate the voltage gain and output impedance for a load resistance of infinity (R_L open). Record your answers in Table 36-4.

10. Connect the circuit using the VMOS transistor of Step 5 with an infinite R_L. Measure and record the voltage gain for an input of 1 V p-p. Also measure and record the MPP value.

11. Measure and record the output impedance using the matched-load method (described in Experiment 34).

Driving a Load

12. Assume the $r_{DS(on)}$ of the VMOS transistor is 5 Ω. Calculate the I_D and V_D in Fig. 36-3. Record your answers in Table 36-5.

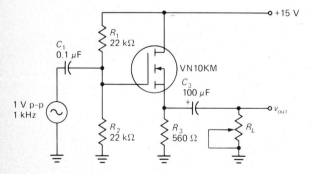

Figure 36-2

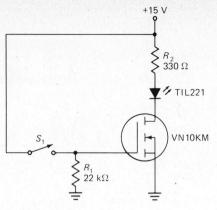

Figure 36-3

13. Connect the circuit. Open and close S_1. The LED should light when S_1 is closed and go out when S_1 is open.

14. Close S_1. Measure I_D and V_D. Record in Table 36-5.

15. Repeat Steps 13 and 14 for other VMOS transistors.

Troubleshooting (Optional)

16. Table 36-6 lists some symptoms for the circuit of Fig. 36-3. Try to figure out the corresponding trouble. Insert the suspected trouble and verify the symptoms. Record each trouble.

Design (Optional)

17. Select a value of current-limiting resistance in Fig. 36-3 that produces an LED current of approximately 20 mA.

18. Connect the circuit with your design value. Measure the LED current. Record your data here:

$R =$ _____

$I =$ _____

Computer (Optional)

19. Enter and run this program:

```
10 R = 1250
20 PRINT "THE RESISTANCE IS"; R;
   "OHMS"
30 STOP
```

20. Write and run a program that calculates the LED current in Fig. 36-3. The inputs are V_{DD}, R_2, V_{LED}, and $r_{DS(on)}$.

DATA FOR EXPERIMENT 36

Table 36-1. Threshold Voltage

VMOS	V_{GSth}
1	
2	
3	

Table 36-2. Transconductance Curve

I_D	V_{GS}
10 μA	
0.5 mA	
1 mA	
2 mA	
3 mA	
4 mA	
5 mA	
6 mA	
7 mA	
8 mA	
9 mA	
10 mA	

Table 36-3. Voltage-Divider Bias

VMOS	Calculated			Measured		
	I_D	V_G	V_S	I_D	V_G	V_S
1						
2						
3						

Table 36-4. Source Follower

Calculated g_m = _____

Calculated A = _____

Calculated r_{out} = _____

Measured A = _____

Measured PP = _____

Measured r_{out} = _____

Table 36-5. Driving a Load

VMOS	Calculated		Measured	
	I_D	V_D	I_D	V_D
1				
2				
3				

Table 36-6. Troubleshooting

S_1	V_D	LED	Trouble
Closed	0 V	out	
Closed	$\cong +14$ V	out	
Open	0 V	on	

QUESTIONS FOR EXPERIMENT 36

1. The threshold voltage of the VMOS was the gate voltage that produced a   (   )
   drain current of:
   (a) 10 μA; (b) 100 μA; (c) 1 mA; (d) 10 mA.
2. When the drain current is approximately 10 mA, the transconductance of the   (   )
   VMOS transistor used in this experiment was closest to:
   (a) 100 μS; (b) 1000 μS; (c) 2500 μS; (d) 30,000 μS.
3. The calculated drain current in Table 36-3 is approximately: (   )
   (a) 1.1 mA; (b) 5.26 mA; (c) 9.82 mA; (d) 12.5 mA.
4. The voltage gain in Table 36-4 is: (   )
   (a) less than 0.5; (b) slightly less than unity; (c) greater than unity;
   (d) around 10.
5. The measured drain voltage of Table 36-5 was: (   )
   (a) large; (b) less than 1 V; (c) equal to supply voltage; (d) unstable.
6. Describe what the VMOS transistor does in Fig. 36-3.

Troubleshooting (Optional)

7. What is the last trouble you recorded in Table 36-6? Why does it produce the given
   symptoms?

8. Suppose the dc voltage across the 560-Ω resistor of Fig. 36-2 is zero. Name three possible troubles.

Design (Optional)

9. What value of resistance did you record in Step 17? How did you arrive at this resistance?

10. Optional. Instructor's question.

EXPERIMENT 37

THE SILICON CONTROLLED RECTIFIER

▼

The silicon controlled rectifier (SCR) acts like a normally off switch. To turn it on, you have to apply a trigger to the gate. Once on, the SCR acts like a closed switch. You can then remove the trigger and the SCR remains closed. The only way to open the SCR is to reduce the supply voltage to a low value near zero.

REQUIRED READING

Chapter 15 (Secs. 15-1 to 15-3) of *Electronic Principles*, 5th ed.

EQUIPMENT

1 power supply: adjustable from approximately 0 to 15 V with current-limiting
1 red LED: TIL221 or equivalent
1 green LED: TIL222 or equivalent
1 zener diode: 1N753
2 transistors: 2N3904, 2N3906
1 SCR: 2N4444
1 op amp: 741C
7 ½-W resistors: 220 Ω, two 330 Ω, 470 Ω, two 1 kΩ, 10 kΩ
1 potentiometer: 1 kΩ
1 VOM (analog or digital multimeter)
1 oscilloscope

PROCEDURE

Transistor Latch

1. The transistor latch of Fig. 37-1 simulates an SCR. Assume the LED of Fig. 37-1 is off. For a V_{CC} of +15 V, calculate the voltage between point A and ground. Record in Table 37-1. Also calculate and record the LED current.

2. Assume the switch of Fig. 37-1 is momentarily closed, then opened. Calculate and record the voltage at point A to ground for a V_{CC} of +15 V. Also calculate and record the LED current.

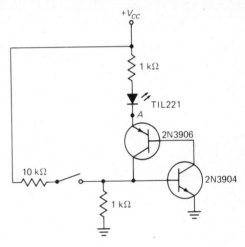

Figure 37-1

3. Connect the circuit with the switch open and a V_{CC} of +15 V.
4. The LED should be out. If not, reduce the supply voltage to zero, then back to +15 V.
5. With the LED off, measure and record the voltage at point A and the LED current.
6. Close the switch. The LED should come on.
7. Open the switch. The LED should stay on.
8. With the LED on, measure and record the voltage at point A. Measure and record the LED current. (When you break the circuit to insert the ammeter, the LED will go out. Close the switch to turn on the LED and measure the current.)
9. With the LED on, reduce the supply voltage until the LED goes off. Then increase the supply voltage and notice that the LED remains off.

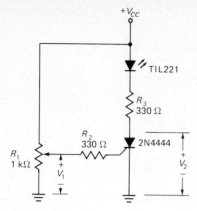

Figure 37-2

10. Close the switch, then open it. The LED should be on.

SCR Circuit

11. In Fig. 37-2, assume that V_{CC} is $+15$ V and that the LED is off. Calculate and record V_2 in Table 37-2. Also calculate and record the LED current.

12. Assume the LED is on. Calculate and record V_2 and I_{LED}.

13. The typical gate trigger current of a 2N4444 is 7 mA (see data sheet in Appendix). The typical gate trigger voltage is 0.75 V. If the LED is off, what is the value of V_1 needed to turn on the LED? Record your answer in Table 37-2.

14. The typical holding current of a 2N4444 is 6 mA. Calculate and record the value of V_{CC} for this holding current.

15. Connect the circuit with R_1 reduced to zero.

16. Adjust V_{CC} to $+15$ V. The LED should be out. (If the LED is on, reduce V_{CC} to zero, then increase it back to $+15$ V. The LED should now be out.) Measure and record V_2 and I_{LED}.

17. Slowly increase V_1 until the LED just comes on. Measure and record V_2 and I_{LED}. Also measure and record V_1.

18. Reduce V_1 to zero. Slowly decrease V_{CC} until the LED just goes out. Measure and record V_{CC}.

DATA FOR EXPERIMENT 37

Table 37-1. Transistor Latch

	Calculated		Measured	
LED	V_A	I_{LED}	V_A	I_{LED}
Off				
On				

Table 37-2. SCR Circuit

LED off:

Calculated V_2 = _____

Calculated I_{LED} = _____

Measured V_2 = _____

Measured I_{LED} = _____

LED on;

Calculated V_2 = _____

Calculated I_{LED} = _____

Measured V_2 = _____

Measured I_{LED} = _____

Triggering:

Calculated V_1 = _____

Measured V_1 = _____

Holding:

Calculated V_{CC} = _____

Measured V_{CC} = _____

QUESTIONS FOR EXPERIMENT 37

1. When the LED of Fig. 37-1 is on, the voltage from point A to ground is ( )
   closest to:
   (a) 0; (b) 3 V; (c) 10 V; (d) V_{CC}.
2. After the LED of Fig. 37-1 comes on, the current through it is approximately: ( )
   (a) 0; (b) 5 mA; (c) 9 mA; (d) 13 mA.
3. When the switch of Fig. 37-1 is closed, the voltage across the lower 1 kΩ is: ( )
   (a) 0; (b) 0.7 V; (c) 1.5 V; (d) 15 V.
4. When the switch of Fig. 37-1 is closed, the current through the 2N3904 is ( )
   closest to:
   (a) 0; (b) 5 mA; (c) 8 mA; (d) 14 mA.
5. After the LED of Fig. 37-2 comes on, the only way to make it go off is to ( )
   (a) reduce V_1 to zero; (b) increase V_1 to 15 V; (c) reduce V_{CC} toward
   zero; (d) increase V_{CC} to 15 V.
6. The calculated V_1 of Table 37-2 needed for triggering is closest to: ( )
   (a) 1 V; (b) 1.9 V; (c) 3.06 V; (d) 4.78 V.

7. Assume a LED voltage of 2 V in Fig. 37-2. The current through the SCR ( )
   when it is conducting is closest to:
   (a) 0;   (b) 20 mA;   (c) 40 mA;   (d) 75 mA.
8. When V_{CC} is 15 V and the LED is off in Fig. 37-2, V_2 is equal to: ( )
   (a) 1 V;   (b) 1.9 V;   (c) 3.06 V;   (d) 15 V.
9. Optional. Instructor's question.

10. Optional. Instructor's question.

EXPERIMENT 38

LOWER CRITICAL FREQUENCIES

▼

At lower frequencies the coupling capacitors no longer look like ac shorts. This implies that some of the ac voltage is dropped across the coupling capacitors. Likewise, the emitter bypass capacitor no longer appears as an ac short when the frequency is low enough. The input coupling capacitor, the output coupling capacitor, and the emitter bypass capacitor each have a critical frequency associated with it. The highest of the critical frequencies is the dominant one because the amplifier response breaks first at this frequency. In this experiment you will build and measure the lower critical frequencies of a CE amplifier. This will allow you to see how coupling and bypass capacitors affect the response of an amplifier at low frequencies.

Up to now, we have limited the troubleshooting to shorted and open components, caused either by defective components or by cold-solder joints and solder bridges. In the production of electronic equipment, you sometimes encounter a component of incorrect value. For instance, in the stuffing of printed-circuit boards, someone may occasionally insert the wrong value for a resistor or capacitor. In this experiment the troubleshooting option introduces components of the wrong size.

REQUIRED READING

Chapter 16 (Secs. 16-1 to 16-4) of *Electronic Principles*, 5th ed.

EQUIPMENT

1 audio generator
1 power supply: 10 V
1 transistor: 2N3904
9 ½-W resistors: 220 Ω, three 1 kΩ, 2.2 kΩ, 3.6 kΩ, 8.2 kΩ, 10 kΩ, 36 kΩ
5 capacitors: 0.1 μF, 1 μF, two 10 μF, 470 μF (16-V rating or better)
1 VOM (analog or digital multimeter)
1 oscilloscope
1 frequency counter

PROCEDURE

Dominant Input Critical Frequency

1. Assume $C_1 = 0.1$ μF, $C_2 = 10$ μF, and $C_3 = 470$ μF in Fig. 38-1. Also assume a typical h_{fe} of 120. Calculate the three lower critical frequencies. Record your answers in Table 38-1.
2. Connect the circuit with the foregoing values.
3. Set the audio generator to 10 kHz with a peak-to-peak value of 20 mV. This is measured from the left end of the 1 kΩ resistor to ground.
4. Look at the output voltage with the oscilloscope. Also measure the output voltage with the VOM.
5. Decrease the input frequency until the output voltage is down to 0.707 of its value at 10 kHz.
6. Check that the source voltage is still 20 mV p-p. If it has changed, readjust it until it again equals 20 mV p-p.

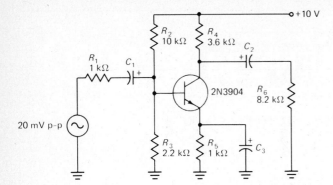

Figure 38-1

7. Repeat Steps 5 and 6 until the source voltage is 20 mV p-p and the output voltage is down to 0.707 of its value at 10 kHz. The frequency is now approximately at the dominant lower critical frequency.

8. Measure the frequency with an electronic frequency counter. (If a counter is not available, use the oscilloscope to measure the period; then calculate the frequency with $f = 1/T$.) Record the critical frequency in Table 38-1.

Dominant Output Critical Frequency

9. Assume $C_1 = 10$ μF, $C_2 = 0.1$ μF, and $C_3 = 470$ μF. Calculate and record the three lower critical frequencies.

10. Repeat Steps 2 to 8.

Dominant Emitter Bypass Critical Frequency

11. Assume $C_1 = 10$ μF, $C_2 = 10$ μF, $C_E = 10$ μF. Calculate and record the three lower critical frequencies.

12. Repeat Steps 2 to 8.

Troubleshooting (Optional)

13. Connect the circuit with $C_1 = 10$ μF, $C_2 = 0.1$ μF, and $C_3 = 470$ μF. Measure the lower critical fre-

quency. It should be in the vicinity of 20 Hz. (Note: All critical frequencies are fairly close to 10 Hz, so the combined effect of the three effects is an amplifier critical frequency near 20 Hz.)

14. Set the input frequency at 10 kHz and the source voltage at 20 mV p-p.

15. As mentioned in the preliminary remarks, sometimes the trouble is a component of incorrect size. In this part of the experiment, the possible troubles are the following: $R_2 = 1$ kΩ, $R_3 = 220$ Ω, $R_4 = 36$ kΩ, and $R_5 = 220$ Ω. Table 38-2 lists representative dc and ac voltages for each of the foregoing troubles. Try to figure out what each trouble is. When you think you have it, insert the trouble in the circuit.

16. Check the dc and ac voltages to verify that you have found the trouble. Record each trouble in Table 38-2.

Design (Optional)

17. Select a value for C_1 that produces an input critical frequency of approximately 2 Hz. Select a value of C_2 to get the same output critical frequency. Select a value of C_3 to produce an emitter bypass critical frequency of approximately 20 Hz. Record the nearest standard capacitance values you have selected (Table 38-3).

18. Connect the circuit with your design values of C_1, C_2, and C_3.

19. Measure and record the dominant lower critical frequency.

Computer (Optional)

20. Write and run a program that calculates the three lower critical frequencies of Fig. 38-1. Use INPUT statements to enter all required data.

DATA FOR EXPERIMENT 38

Table 38-1. Lower Critical Frequencies

		Calculated		Measured
Circuit	f_{in}	f_{out}	f_E	f_c
1				
2				
3				

Table 38-2. Troubleshooting

	DC Voltages			AC Voltages			
V_B	V_E	V_C	v_b	v_e	v_c	Trouble
1.8 V	1.1 V	6 V	10.4 mV	0	1.14 V	None
0.24 V	0	10 V	3.6 mV	0	0	
1.2 V	0.57 V	0.58 V	1.5 mV	0	0	
5 V	4.3 V	4.31 V	0	0	0	
1.38 V	0.7 V	0.76 V	1 mV	0	2 mV	

Table 38-3. Design

$C_1 =$ _____

$C_2 =$ _____

$C_3 =$ _____

Measured $f_c =$ _____

QUESTIONS FOR EXPERIMENT 38

1. Of the three lower critical frequencies, the dominant or most important one ( )
   is the:
   (a) input critical frequency;   (b) output critical frequency;   (c) emitter
   bypass critical frequency;   (d) highest of the three.

2. Which of the following usually has the largest capacitance in an amplifier ( )
   stage like Fig. 38-1?
   (a) Input coupling capacitor;   (b) output coupling capacitor;   (c) emitter
   bypass capacitor.

3. If any capacitor increases by a factor of 10, the critical frequency associated ( )
   with this capacitor:
   (a) decreases by a factor of 10;   (b) increases by a factor of 10;   (c) stays
   the same;   (d) none of the foregoing.

4. When C_3 is 10 μF, the calculated critical frequency for the emitter bypass ( )
   capacitor is closest to:
   (a) 7.63 Hz;   (b) 12.1 Hz;   (c) 135 Hz;   (d) 569 Hz.

5. If all three lower critical frequencies are equal to 10 Hz, the amplifier critical   (   )
   frequency will be:
   (a) less than 10 Hz;   (b) 10 Hz;   (c) more than 10 Hz;   (d) 100 Hz.
6. Briefly explain why there are lower critical frequencies in the circuit of Fig. 38-1.

Troubleshooting (Optional)

7. In the second trouble of Table 38-2, the dc collector voltage is 10 V. What trouble
   did you locate here? Explain why the trouble forced the transistor into cutoff.

8. An amplifier like Fig. 38-1 has a lower critical frequency approximately 10 times
   higher than it should be. All dc voltages are normal. Describe how you would use an
   oscilloscope to isolate the fault.

Design (Optional)

9. What critical frequency did you measure? Give some reasons why the measured critical
   frequency differed from 20 Hz.

10. Optional. Instructor's question.

EXPERIMENT 39

UPPER CRITICAL FREQUENCIES

▼

At higher frequencies the internal capacitances of an active device like a JFET or bipolar transistor produce upper critical frequencies. With a JFET amplifier, there are two upper critical frequencies: the gate critical frequency and the drain critical frequency. With a bipolar amplifier the two critical frequencies are the base critical frequency and the collector critical frequency. In this experiment you will build a JFET amplifier and measure its upper critical frequencies.

REQUIRED READING

Chapter 16 (Secs. 16-5 to 16-13) of *Electronic Principles*, 5th ed.

EQUIPMENT

1 audio generator
1 power supply: 15 V
1 JFET: MPF102 (or any *n*-channel JFET with an I_{DSS} greater than 2 mA)
5 ½-W resistors: 100 Ω, two 2.2 kΩ, 10 kΩ, 24 kΩ
7 capacitors: three 1000 pF, 0.0027 μF, 0.01 μF, 0.033 μF, 100 μF (16-V rating or better)
1 VOM (analog or digital multimeter)
1 oscilloscope
1 frequency counter

PROCEDURE

JFET Amplifier

1. To simplify this experiment, we add large capacitors to the JFET as shown in Fig. 39-1. This brings the upper critical frequency down low enough to measure easily. Because internal FET capacitances are normally less than 10 pF, the effective values to use in this experiment become:

$$C_{gs} = C_{gd} = C_{ds} \cong 1000 \text{ pF}$$

2. In Fig. 39-1, calculate the approximate value of r_G in the gate bypass network. Using a g_m of 2000 μS, calculate the capacitance C_G for the gate bypass network. Record r_G and C_G in Table 39-1.
3. Calculate the critical bypass frequency of the gate network and record the value in Table 39-1.
4. In Fig. 39-1, what is the drain resistance r_D? Record the answer in Table 39-2.
5. Calculate the capacitance C_D and the critical frequency f_D of the drain bypass network. Record the values in Table 39-2.
6. Connect the JFET amplifier of Fig. 39-1.
7. Adjust the audio generator to 100 Hz with a peak-to-peak input voltage of 200 mV.
8. Look at the output voltage with an oscilloscope. It should be a sine wave with a peak-to-peak voltage in the vicinity of 1 V.
9. Measure the rms output voltage with a VOM.
10. The gate bypass network is dominant (see Tables 39-1 and 39-2). The critical frequency of the gate lag network should be in the vicinity of the f_G in Table 39-1. Find the actual frequency by locating the frequency where the voltage gain is down to 0.707 of its value at 100 Hz. Record this critical frequency in Table 39-3.
11. Short out the 10-kΩ resistor. This removes the gate bypass network. Find the critical frequency of the drain bypass network by increasing the frequency until the output voltage is down to 0.707 of its value at 100 Hz. Record the value of f_D for the drain bypass network in Table 39-3.

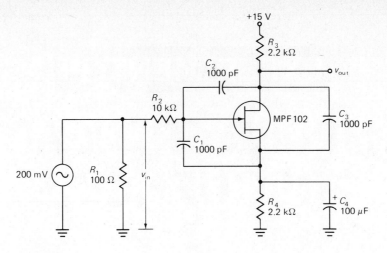

Figure 39-1

Troubleshooting (Optional)

12. Suppose the circuit of Fig. 39-1 has an upper critical frequency in the vicinity of 1 kHz. Assume this trouble is being caused by a component of incorrect size. Try to locate four troubles that produce this symptom. Insert each trouble in the circuit and verify that the critical frequency is in the vicinity of 1 kHz. Record each trouble in Table 39-4.

Design (Optional)

13. Select a value for C_2 that produces an upper critical frequency of approximately 5 kHz. Record the nearest standard capacitance in Table 39-5.

14. Connect the circuit with your design value of C_2.

15. Measure and record the upper critical frequency.

Computer (Optional)

16. Write and run a program that calculates the gate and drain critical frequencies in Fig. 39-1. Use INPUT statements to enter all required data.

DATA FOR EXPERIMENT 39

Table 39-1. Gate Bypass Network

$r_G =$ _____

$C_G =$ _____

$f_G =$ _____

Table 39-2. Drain Bypass Network

$r_D =$ _____

$C_D =$ _____

$f_D =$ _____

Table 39-3. Measured Critical Frequencies

$f_G =$ _____

$f_D =$ _____

Table 39-4. Troubleshooting

Number	Trouble
1	
2	
3	
4	

Table 39-5. Design

$C_2 =$ _____

$f_c =$ _____

QUESTIONS FOR EXPERIMENT 39

1. Of the two upper critical frequencies, the dominant or most important one   (   )
   is the:
   (a) lower of the two;   (b) drain critical frequency;   (c) source bypass critical frequency;   (d) higher of the two.
2. The gate critical frequency was closest to: (   )
   (a) 2.49 kHz;   (b) 36.2 kHz;   (c) 45 kHz;   (d) 67 kHz.
3. In this experiment which of the following was the dominant critical frequency?   (   )
   (a) source critical frequency;   (b) gate critical frequency;   (c) drain critical frequency;   (d) none of the foregoing.
4. If the voltage gain is 4.4, the Miller input capacitance of Fig. 39-1 is: (   )
   (a) 1000 pF;   (b) 2000 pF;   (c) 4400 pF;   (d) 5400 pF.

5. If the gate critical frequency and the drain critical frequency are each equal ( )
to 100 kHz, the amplifier critical frequency will be:
(**a**) less than 100 kHz; (**b**) 100 kHz; (**c**) more than 100 kHz;
(**d**) 1 MHz.

6. Briefly explain why there are upper critical frequencies in the circuit of Fig. 39-1.

Troubleshooting (Optional)

7. Suppose C_4 opens in Fig. 39-1. What effect will this have on the amplifier response?

8. Somebody substitutes another JFET for the MPF102 of Fig. 39-1. If the upper critical frequency of the amplifier is only half of what it is with the MPF102, what do you think the trouble is?

Design (Optional)

9. What was the critical frequency you measured with your design value of C_2? Give some reasons why this differs from the theoretical value of 5 kHz.

10. Optional. Instructor's question.

EXPERIMENT 40

DECIBELS AND RISETIME

▼

Decibles can be used to specify power gain or voltage gain. With cascaded stages, the overall decibel gain is the sum of the individual decibel gains of the stages. Decibels are also useful when drawing Bode plots of amplifier response. The decibel gain is down 3 dB at the cutoff frequency of a coupling or bypass network. Beyond cutoff, the gain rolls off at a rate of 6 dB per octave, equivalent to 20 dB per decade for each coupling or bypass network.

A quick way to find the cutoff frequency of an amplifier is to use a square-wave input and measure the risetime of the output voltage. For an amplifier with one dominant bypass network, the upper critical frequency equals 0.35 divided by the risetime.

REQUIRED READING

Chapter 16 (Secs. 16-9 to 16-11 and 16-18) of *Electronic Principles*, 5th ed.

EQUIPMENT

- 1 sine/square generator (or two generators: one sine wave and one square wave)
- 1 decade resistance box
- 11 ½-W resistors: four 470 Ω, two 1 kΩ, 2.2 kΩ, 3.6 kΩ, 10 kΩ, 22 kΩ, 100 kΩ
- 1 transistor: 2N3904
- 4 capacitors: 0.01 μF, 0.022 μF, 10 μF, 470 μF
- 1 ac voltmeter with decibel scale
- 1 oscilloscope

PROCEDURE

Reading Decibel Scale

1. Figure 40-1a shows a voltage divider. For each value of R given in Table 40-1, calculate the decibel voltage gain. Round off your answers to the nearest decibel and record the values in Table 40-1.
2. Connect the circuit of Fig. 40-1a using an R of 240 kΩ.

3. Put the ac voltmeter on the 1-V range. Adjust the audio generator to get a v_{in} reading of 0 dB.
4. Measure v_{out} using the dB scale. This reading is the decibel voltage gain. Record the value in Table 40-1 (round off to the nearest decibel).
5. Repeat Steps 3 and 4 for the other R values listed in Table 40-1.

Decibels Add

6. Figure 40-1b shows three cascaded voltage dividers. Let $A_1 = v_2/v_1$, $A_2 = v_3/v_2$, and $A_3 = v_4/v_3$. Round off 470 to 500 and calculate the decibel voltage gain A_1', A_2', and A_3'. Record in Table 40-2.
7. Add these decibel voltage gains to get the total decibel voltage gain. Record this value as A' in Table 40-2.
8. Connect the circuit of Fig. 40-1b.
9. Set v_1 at 0 dB on the 1-V range of the ac voltmeter.
10. Read the values of v_2, v_3, and v_4 rounded off to the nearest decibel and record in Table 40-3.

Bypass Circuit

11. Calculate the critical frequency of the bypass network shown in Fig. 40-2. Record this value in Table 40-4.

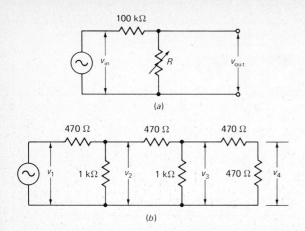

(a)

(b)

Figure 40-1

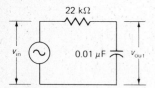

Figure 40-2

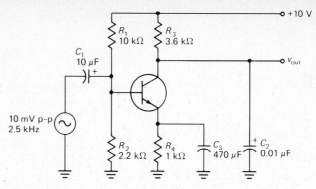

Figure 40-3

12. Assume a square wave is the input signal in Fig. 40-2. Calculate and record the risetime of the output.

13. Connect the circuit of Fig. 40-2. Adjust the generator to deliver a sine wave. Set the frequency to the value of f_c in Table 40-4 and the signal level to a v_{in} of 0 dB on the 1-V range of the ac voltmeter.

14. Read the output voltage in decibels. Adjust the frequency slightly until it is down 3 dB from the input. Record the measured critical frequency in Table 40-5.

15. Put a 300-Hz square wave into the lag network. Look at the input signal with the oscilloscope and set the level at 1 V p-p.

16. Measure the risetime of the output signal. Record T_R in Table 40-5.

Amplifier Critical Frequency

17. In Fig. 40-3, you may assume the base critical frequency is much higher than the collector critical frequency. Furthermore, you can ignore the internal collector capacitance. Calculate and record the critical frequency of the amplifier (Table 40-6).

18. Connect the circuit. Measure and record the risetime. Calculate the critical frequency using the measured risetime. Record this as the experimental critical frequency.

Troubleshooting (Optional)

19. Table 40-7 lists some troubles. Try to figure out what can cause the trouble. When you think you have it, insert the trouble and verify the symptoms. Record each trouble you locate.

Design (Optional)

20. Select a value of C_2 to get a risetime of approximately 375 μs. Record the nearest standard capacitance in Table 40-8.

21. Connect the circuit with your design value of C_2. Measure and record the risetime of the output voltage.

Computer (Optional)

22. Write and run a program that prints out the cutoff frequency for all risetimes between 1 and 100 μs (1, 2, 3, . . . , 100). Your program should use a FOR . . . NEXT statement with 100 steps.

DATA FOR EXPERIMENT 40

Table 40-1. Decibels

R	Calculated A'	Measured A'
240 kΩ		
100 kΩ		
46 kΩ		
11.1 kΩ		
1 kΩ		

Table 40-2. Adding Decibels

$A_1' =$

$A_2' =$

$A_3' =$

$A' =$

Table 40-3. Measurements

$v_1 = 0$ dB

$v_2 =$

$v_3 =$

$v_4 =$

Table 40-4. Bypass Calculations

$f_c =$

$T_R =$

Table 40-5. Bypass Measurements

$f_c =$

$T_R =$

Table 40-6. Amplifier Risetime

Calculated critical frequency =

Measured risetime =

Experimental critical frequency =

Table 40-7. Troubleshooting

Symptoms	Trouble
$A = -3.6$	
$T_R = 175 \ \mu s$, dc ok	
DC ok, no ac output	

Table 40-8. Design

$C_2 =$	
$T_R =$	

QUESTIONS FOR EXPERIMENT 40

1. Negative values of decibel voltage gain mean that the ordinary voltage gain ( )
   is:
   (a) negative; (b) phase inverted; (c) less than unity; (d) less than
   zero.
2. With cascaded stages, the decibel voltage gain of each stage is added to get ( )
   the:
   (a) antilog of the voltage gain; (b) total ordinary voltage gain; (c) total
   decibel voltage gain; (d) cutoff frequency.
3. The calculated risetime in Table 40-4 is approximately: ( )
   (a) 123 μs; (b) 484 μs; (c) 678 μs; (d) 1000 μs.
4. In Table 40-6, the risetime is closest to: ( )
   (a) 10 μs; (b) 23 μs; (c) 36 μs; (d) 79 μs.
5. To calculate the risetime, you can divide 0.35 by the ( )
   (a) risetime; (b) voltage gain; (c) supply voltage; (d) upper critical
   frequency.
6. The neat relation, $f_c = 0.35/T_R$, can be used only if the amplifier satisfies a certain
   condition. What is this condition?

Troubleshooting (Optional)

7. What was the first trouble you recorded in Table 40-7? Why does this trouble produce
   such a low voltage gain?

8. If someone uses 0.001 μF for C_2 in Fig. 40-3, what will happen to the risetime?

Design (Optional)

9. If you are trying to design an amplifier to work at high frequencies, do you want a small or a large risetime? Explain your answer.

10. Optional. Instructor's question.

EXPERIMENT 41

THE DIFFERENTIAL AMPLIFIER

▼

The differential amplifier is the direct-coupled input stage of the typical op amp. The most common form of a diff amp is the double-ended input and single-ended output circuit. Some of the important characteristics of a diff amp are the input offset current, input bias current, input offset voltage, and common-mode rejection ratio. In this experiment you will build a diff amp and measure the foregoing quantities.

REQUIRED READING

Chapter 17 (Secs. 17-1 to 17-6) of *Electronic Principles*, 5th ed.

EQUIPMENT

1 audio generator
2 power supplies ±15 V
10 ½-W resistors: two 22 Ω, two 100 Ω, two 1.5 kΩ, two 4.7 kΩ, two 10 kΩ (5% tolerance)
2 transistors: 2N3904
1 capacitor: 0.47 μF
1 VOM (analog or digital multimeter)
1 oscilloscope

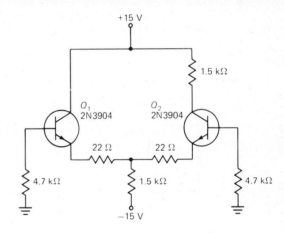

Figure 41-1

PROCEDURE

Tail Current and Base Currents

1. Notice the pair of swamping resistors (22 Ω) in Fig. 41-1. These have to be included in this experiment to improve the match between the discrete transistors. In Fig. 41-1, you may assume the typical h_{FE} is 200. Calculate the approximate tail current. Record in Table 41-1. Also calculate and record the base current in each transistor.
2. Connect the circuit of Fig. 41-1.
3. Measure and record the tail current.
4. Use the VOM as an ammeter to measure the base current in each transistor. If your VOM is not sensitive enough to measure microampere currents, then use the oscilloscope on dc input to measure the voltage across each base resistor and calculate the

base current. Record the base currents in Table 41-1.

Input Offset and Bias Currents

5. With the calculated data of Table 41-1, calculate the values of input offset current and input bias current. Record your theoretical answers in Table 41-2.
6. With the measured data of Table 41-1, calculate the values of $I_{in(off)}$ and $I_{in(bias)}$. Record your experimental answers in Table 41-2.

Output Offset Voltage

7. In Fig. 41-2, assume the base of Q_1 is grounded by a jumper wire. If both transistors are identical and

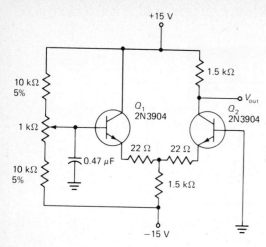

Figure 41-2

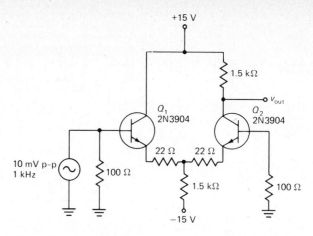

Figure 41-3

all components have the values shown, then the dc output voltage would have a value of approximately $+7.85$ V. For this part of the experiment, any deviation from $+7.85$ V is called output offset voltage, designated $V_{out(off)}$.

8. Connect the circuit of Fig. 41-2. Ground the base of Q_1 with a jumper wire. Measure the dc output voltage. Calculate the output offset voltage and record $V_{out(off)}$ in Table 41-3.

9. Remove the ground from the Q_1 base. Adjust the potentiometer until the output voltage is $+7.85$ V.

10. Measure the base voltage of Q_1. Record in Table 41-3 as $V_{in(off)}$.

Differential Voltage Gain

11. Because of the swamping resistors in Fig. 41-3, the differential voltage gain is given by $R_C/2(r_E + r_e')$. Calculate and record A in Table 41-4.

12. Connect the circuit. Set the audio generator at 1 kHz with a signal level of 10 mV p-p.

13. Measure the output voltage. Calculate and record the experimental value of A.

Common-Mode Voltage Gain

14. Calculate the common-mode voltage gain of Fig. 41-3. Record A_{CM} in Table 41-4.

15. Put a jumper wire between the bases of your built-up circuit.

16. Increase the signal level until the output voltage is 0.5 V p-p.

17. Measure the peak-to-peak input voltage. Calculate and record the experimental value of A_{CM}.

Common-Mode Rejection Ratio

18. Calculate and record the theoretical value of CMRR using the calculated data of Table 41-4.

19. Calculate and record the experimental value of CMRR using the experimental data of Table 41-4.

Troubleshooting (Optional)

20. In this part of the experiment, a collector-emitter short means all three transistor terminals are shorted together. A collector-emitter open means the transistor is removed from the circuit.

21. In Fig. 41-3, estimate the dc output voltage for each trouble listed in Table 41-5.

22. Insert each trouble, measure and record the dc voltages of Table 41-5.

Design (Optional)

23. Select resistance values in Fig. 41-3 to get a tail current of 3 mA and a dc output voltage of $+7.5$ V. Record the nearest standard values in Table 41-6.

24. Connect the circuit with your design values. Measure and record the tail current and dc output voltage.

Computer (Optional)

25. Enter and run this program:

```
10 A$ = "ENTER SUPPLY VOLTAGE"
20 B$ = "VEE": C$ = "VCC"
30 PRINT A$;B$: INPUT VEE
40 PRINT A$;C$: INPUT VCC
50 PRINT "VEE = "; VEE
60 PRINT "VCC = "; VCC
70 END
```

26. Write and run a program that calculates the differential voltage gain, common-mode voltage gain, and common-mode rejection ratio of Fig. 41-3. The inputs are the supply voltages and resistances of the circuit.

DATA FOR EXPERIMENT 41

Table 41-1. Tail and Base Currents

Calculated	Measured
$I_T =$ _____	$I_T =$ _____
$I_{B1} =$ _____	$I_{B1} =$ _____
$I_{B2} =$ _____	$I_{B2} =$ _____

Table 41-2. Input Offset and Bias Currents

Theoretical	Experimental
$I_{in(off)} =$ _____	$I_{in(off)} =$ _____
$I_{in(bias)} =$ _____	$I_{in(bias)} =$ _____

Table 41-3. Input and Output Offset Voltages

$V_{out(off)} =$

$V_{in(off)} =$

Table 41-4. Voltage Gains and CMRR

Calculated	Experimental
$A =$ _____	$A =$ _____
$A_{CM} =$ _____	$A_{CM} =$ _____
CMRR $=$ _____	CMRR $=$ _____

Table 41-5. Troubleshooting

Trouble	Estimated V_{out}	Measured V_{out}
Q_1 CE short		
Q_1 CE open		
Q_2 CE short		
Q_2 CE open		

Table 41-6. Design

$R_E =$

$R_C =$

$I_T =$

$V_{C2} =$

QUESTIONS FOR EXPERIMENT 41

1. The tail current of Table 41-1 is closest to: ( )
   (a) 1 μA;   (b) 23.8 μA;   (c) 47.6 μA;   (d) 9.53 mA.

2. The calculated base current of Fig. 41-1 is approximately;   ( )
   (a) 1 μA;   (b) 23.8 μA;   (c) 47.6 μA;   (d) 9.53 mA.

3. The calculated input bias current of Fig. 41-1 is approximately: ( )
   (a) 1 μA;   (b) 23.8 μA;   (c) 47.6 μA;   (d) 9.53 mA.

4. The input offset voltage is the input voltage that removes the: ( )
   (a) tail current;   (b) dc output voltage;   (c) output offset voltage;
   (d) supply voltage.

5. The CMRR of Table 41-4 is closest to: ( )
   (a) 0.5;   (b) 27.5;   (c) 55;   (d) 123.

6. Why is a high CMRR an advantage with a diff amp?

Troubleshooting (Optional)

7. In Fig. 41-3, somebody mistakenly uses 150 Ω instead of 1.5 kΩ for the tail resistor. What are some of the dc and ac symptoms you can expect?

8. You are troubleshooting the circuit of Fig. 41-3. What should an oscilloscope display for the ac voltage between the 22-Ω resistor and ground?

Design (Optional)

9. What value did you use for R_E and R_C in your design? What is the new value of CMRR?

10. Optional. Instructor's question.

184

EXPERIMENT 42

THE OPERATIONAL AMPLIFIER

▼

An operational amplifier, or op amp, is a high-gain dc amplifier usable from 0 to over 1 MHz (typical). By connecting external resistors to an op amp, you can adjust the voltage gain and bandwidth to your requirements. Whether troubleshooting or designing, you have to be familiar with the characteristics of an op amp. These include the input offset current, input bias current, input offset voltage, common-mode rejection ratio, MPP value, short-circuit output current, slew rate, and power bandwidth. In this experiment you will connect and test a basic op-amp circuit.

REQUIRED READING

Chapter 18 (Secs. 18-1 to 18-6) of *Electronic Principles*, 5th ed.

EQUIPMENT

1 audio generator
2 power supplies: ±15 V
8 ½-W resistors: two 100 Ω, 1 kΩ, two 10 kΩ, 100 kΩ, two 220 kΩ
3 op amps: 741C
2 capacitors: 0.47 μF
1 VOM (analog or digital multimeter)
1 oscilloscope

PROCEDURE

Input Offset and Bias Currents

1. The 741C has a typical $I_{in(bias)}$ of 80 nA. Assume that this is the base current in each 220-kΩ resistor of Fig. 42-1. Calculate dc voltages at the noninverting and inverting inputs. Record in Table 42-1.
2. Connect the circuit of Fig. 42-1.
3. Measure the dc voltage at the noninverting input. Record in Table 42-1.
4. Measure and record the inverting input voltage.
5. Repeat Steps 1 to 4 for the other 741Cs.
6. With the measured data of Table 42-1, calculate the

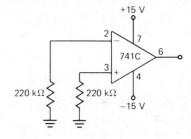

Figure 42-1

base currents, then the values of $I_{in(off)}$ and $I_{in(bias)}$. Record your answers in Table 42-2.

Output Offset Voltage

7. Connect the circuit of Fig. 42-2. Note: Bypass capacitors are used on each supply voltage to prevent oscillations, discussed in Chap. 22 of your textbook. These capacitors should be connected as close to the IC as possible.
8. Measure the dc output voltage. Record this value as $V_{out(off)}$ in Table 42-3.
9. Repeat Step 8 for the other 741Cs.
10. With the resistors shown in Fig. 42-2, the circuit has a voltage gain of 1000. Calculate the input offset voltage with

$$V_{in(off)} = \frac{V_{out(off)}}{1000}$$

Record your results in Table 42-3.

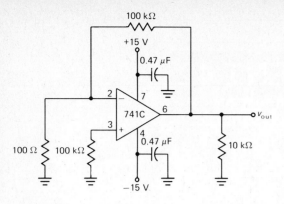

Figure 42-2

Maximum Output Current

11. Disconnect the right end of the 100-kΩ resistor from the output.
12. Connect the right end of the 100-kΩ resistor to the +15 V. This will apply approximately 15 mV to the inverting input, more than enough to saturate the op amp.
13. Replace the 10-kΩ load resistor by a VOM used as an ammeter. Since the ammeter has a very low resistance, it indicates the short-circuit output current.
14. Read and record I_{max} in Table 42-3.
15. Repeat Step 14 for the other 741Cs.

Slew Rate

16. Connect the circuit of Fig. 42-3 with an R_2 of 100 kΩ.
17. Use the oscilloscope (time base around 20 μs/cm) to look at the output of the op amp. Set the audio generator at 5 kHz. Adjust the signal level to get

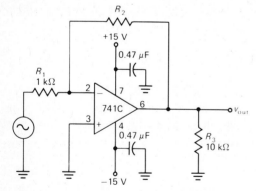

Figure 42-3

hard clipping on both peaks of the output signal (overdrive condition).
18. Measure the voltage change and the time change of the waveform. Calculate and record the slew rate in Table 42-4.
19. Repeat Step 18 for the other 741Cs.

Power Bandwidth

20. Change R_2 to 10 kΩ. Set the ac generator at 1 kHz. Adjust the signal level to get 20 V p-p out of the op amp.
21. Increase the frequency from 1 to 20 kHz and watch the waveform. Somewhere above 8 kHz, slew-rate distortion will become evident because the waveform will appear triangular and the amplitude will decrease.
22. Record the approximate (ballpark) frequency where slew-rate distortion begins (Table 42-4).
23. Repeat Steps 20 to 22 for the other 741Cs.

MPP Value

24. Set the ac generator at 1 kHz. Increase the signal level until clipping just starts on either peak.
25. Record the MPP of all transistors in Table 42-4.

Troubleshooting (Optional)

26. Measure the dc and ac output voltage for each trouble listed in Table 42-5.
27. Record your data in Table 42-5.

Design (Optional)

28. As will be derived in Chap. 20 of your textbook, the voltage gain of a circuit like Fig. 42-3 is equal to R_2/R_1. Select a value of R_2 to get a voltage gain of 68.
29. Replace R_2 by your design value. Measure the voltage gain.
30. Record your design value and the measured voltage gain in Table 42-6.

Computer (Optional)

31. Write and run a program that calculates the power bandwidth of a typical 741C for peak voltages of 1 to 10 V (1, 2, 3, . . . , 10). Your program should use a FOR . . . NEXT statement with 10 steps.

DATA FOR EXPERIMENT 42

Table 42-1. Input Voltages

Op Amp	Calculated		Measured	
	v_1	v_2	v_1	v_2
1				
2				
3				

Table 42-2. Input Offset and Bias Currents

Op Amp	$I_{in(off)}$	$I_{in(bias)}$
1		
2		
3		

Table 42-3. Input and Output Offset Voltages

Op Amp	$V_{out(off)}$	$V_{in(off)}$	I_{max}
1			
2			
3			

Table 42-4. Slew Rate, Power Bandwidth, and MPP Value

Op Amp	S_R	f_{max}	MPP
1			
2			
3			

Table 42-5. Troubleshooting

Trouble	DC Output Voltage	AC Output Voltage
No +15 V		
No −15 V		
Pin 2 shorted to GND		

Table 42-6. Design

$R_2 =$

$A =$

QUESTIONS FOR EXPERIMENT 42

1. The calculated dc voltages in Table 42-1 are approximately: ( )
   (a) 1 mV; (b) 5.6 mV; (c) 12.3 mV; (d) 17.6 mV.
2. The input bias current of Table 42-2 is closest to: ( )
   (a) 1 nA; (b) 80 nA; (c) 2 mA; (d) 25 mA.
3. The short-circuit currents of Table 42-3 are closest to: ( )
   (a) 1 nA; (b) 80 nA; (c) 2 mA; (d) 25 mA.
4. When the input frequency was much higher than the f_{max} of Table 42-4, the   ( )
   output looked:
   (a) sinusoidal; (b) triangular; (c) square; (d) undistorted.
5. The MPP value of Table 42-4 is closest to: ( )
   (a) 5 mV; (b) 15 V; (c) 25 V; (d) 30 V.
6. Explain the meaning of the input offset current and input bias current.

Troubleshooting (Optional)

7. Explain the meaning of the input offset voltage.

8. Describe how you measured the slew rate in this experiment.

Design (Optional)

9. What value did you use for R_2? Why?

10. Optional. Instructor's question.

188

EXPERIMENT 43

NONINVERTING VOLTAGE FEEDBACK

▼

There are four basic types of negative feedback, depending on which input is used and which output quantity is sampled. Noninverting voltage feedback results in an almost perfect voltage amplifier, one with high input impedance, low output impedance, and stable voltage gain. The negative feedback also reduces nonlinear distortion and output offset voltage.

In this experiment you will work with noninverting voltage feedback. First, you will see how accurate the formula for closed-loop voltage gain is. Second, you will see how stable the voltage gain is for different op amps. Third, you will calculate and measure output offset voltages for different feedback resistors. Also included are troubleshooting, design, and computer options.

REQUIRED READING

Chapter 19 (Secs. 19-1 to 19-4) of *Electronic Principles*, 5th ed.

EQUIPMENT

1 audio generator
2 power supplies: ± 15 V
9 $\frac{1}{2}$-W resistors: two 1 kΩ, two 10 kΩ, 22 kΩ, 33 kΩ, 47 kΩ, 68 kΩ, 100 kΩ
3 op amps: 741C
2 capacitors: 0.47 μF
1 VOM (analog or digital multimeter)
1 oscilloscope

PROCEDURE

Voltage Amplifier

1. In Fig. 43-1, assume R_1 equals 10 kΩ. Calculate the closed-loop voltage gain. Record A_{CL} in Table 43-1.
2. Repeat Step 1 for the other values of R_1 shown in Table 43-1.
3. Connect the voltage amplifier of Fig. 43-1 with R_1 equal to 10 kΩ. Set the audio generator to 1 kHz at 100 mV p-p. Measure v_{out}. Calculate the closed-loop

voltage gain with $A_{CL} = v_{out}/v_{in}$. Record this as the measured A_{CL}.
4. Repeat Step 3 for the other values of R_1 listed in Table 43-1.

Stable Voltage Gain

5. Assume R_1 is 33 kΩ in Fig. 43-1. Calculate and record the closed-loop voltage gain (Table 43-2).
6. Connect the circuit with R_1 equal to 33 kΩ. Measure

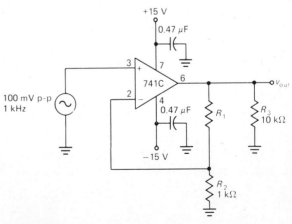

Figure 43-1

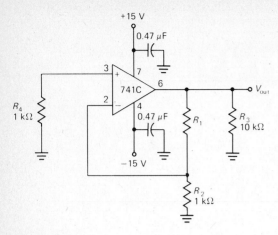

Figure 43-2

v_{out} and calculate A_{CL}. Record this measured value in Table 43-2.

7. Repeat Steps 5 and 6 for the other 741Cs.

Output Offset Voltage

8. Differences in the V_{BE} values and the input base currents imply there is a dc input offset voltage in Fig. 43-2. Assume the total input offset voltage is 2 mV. Calculate and record the output offset voltage for each value of R_1 listed in Table 43-3.

9. Connect the circuit. Measure and record the output offset voltage for each value of R_1. (Even though your measured values may differ considerably from calculated values, the output offset voltage should increase with an increase in R_1.

Troubleshooting (Optional)

10. Assume R_1 is 100 kΩ in Fig. 43-1. For each trouble listed in Table 43-4, estimate the dc and peak-to-peak ac output voltage. Record your estimates in Table 43-4.

11. Connect the circuit with R_1 equal to 100 kΩ. Insert each trouble into the circuit. Measure and record the dc and ac output voltages.

Design (Optional)

12. Select a value of R_1 in Fig. 43-1 to set up a closed-loop voltage gain of 40.

13. Connect the circuit with your design value of R_1. Measure the closed-loop voltage gain. Record your R_1 and A_{CL} in Table 43-5.

Computer (Optional)

14. Write and run a program that calculates the closed-loop voltage gain, input impedance, and output impedance of Fig. 43-1. Use INPUT statements to enter any data needed.

DATA FOR EXPERIMENT 43

Table 43-1. Closed-Loop Voltage Gain

R_1	Calculated A_{CL}	Measured A_{CL}
10 kΩ		
22 kΩ		
47 kΩ		
68 kΩ		
100 kΩ		

Table 43-2. Stable Voltage Gain

Op Amp	Calculated A_{CL}	Measured A_{CL}
1		
2		
3		

Table 43-3. Closed-Loop Output Offset Voltage

R_1	Calculated $V_{oo(CL)}$	Measured $V_{oo(CL)}$
10 kΩ		
22 kΩ		
47 kΩ		
68 kΩ		
100 kΩ		

Table 43-4. Troubleshooting

Trouble	Estimated V_{out}	Estimated v_{out}	Measured V_{out}	Measured v_{out}
R_1 short				
R_1 open				
R_2 short				
R_2 open				

Table 43-5. Design

$R_1 =$

$A_{CL} =$

QUESTIONS FOR EXPERIMENT 43

1. The calculated and measured A_{CL} of Table 43-1 were: ( )
   (a) extremely large; (b) very small; (c) close in value; (d) unpredictable.

2. The measured A_{CL} of Table 43-2 for all three 741Cs was: ( )
   (a) extremely large; (b) very small; (c) almost constant; (d) quite variable.

3. When R increases in Table 43-3, the closed-loop voltage gain increases and ( )
   the output offset voltage:
   (a) decreases; (b) increases; (c) stays the same; (d) none of the foregoing.

4. The closed-loop voltage gain of an amplifier with noninverting voltage feed- ( )
   back is as stable as the:
   (a) supply voltage; (b) gain of the 741C; (c) load resistor;
   (d) feedback resistors.

5. If the input bias current is 80 nA in Fig. 43-2, the dc voltage across R_4 is: ( )
   (a) 80 μV; (b) 800 μV; (c) 2 mV; (d) −15 V.

                                                                                ( )

6. What is the ac voltage at the inverting input of Fig. 43-1? Why?

Troubleshooting (Optional)

7. When R_1 is open or R_2 is shorted in Fig. 43-1, you get a clipped output with a peak-to-peak value around 28 V. Explain why this happens.

8. When R_1 is shorted or R_2 is open in Fig. 43-1, what does the closed-loop voltage gain equal? What is the name for this kind of circuit?

Design (Optional)

9. You are designing a voltage amplifier like Fig. 43-1. If you want to get a voltage gain accurate to within 2 percent, what would you specify in your design?

10. Optional. Instructor's question.

EXPERIMENT 44

NEGATIVE FEEDBACK

▼

Always remember that there are four distinct types of negative feedback. Each type has different characteristics. Noninverting voltage feedback results in a voltage amplifier. Noninverting current feedback leads to a voltage-to-current converter. Inverting voltage feedback results in a current-to-voltage converter. Inverting current feedback leads to a current amplifier.

All four types of negative feedback reduce nonlinear distortion and output offset voltage. The noninverting types increase the input impedance, while the inverting types decrease it. The voltage feedback types decrease the output impedance, while the current feedback types increase the output impedance.

In this experiment, you will connect all four types of negative-feedback circuits using dc input and output voltages and currents.

REQUIRED READING

Chapter 19 (Secs. 19-5, 19-9, and 19-10) of *Electronic Principles*, 5th ed.

EQUIPMENT

1 audio generator
2 power supplies: ±15 V
6 ½-W resistors: two 1 kΩ, 2 kΩ, two 10 kΩ, 18 kΩ
1 potentiometer: 1 kΩ
1 op amp: 741C
2 capacitors: 0.47 μF
2 VOMs: If two VOMs are not available, you can run the experiment with only one

PROCEDURE

Voltage Amplifier

1. For each dc input voltage listed in Table 44-1, calculate the dc output voltage of Fig. 44-1. Record your answers.
2. Connect the circuit. Use one VOM on the input side and one on the output side. (If you don't have two VOMs, you will have to measure the input voltage first, then the output voltage.)

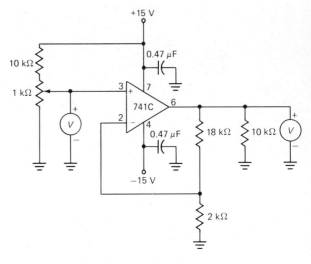

Figure 44-1

3. Adjust the potentiometer to get each dc input voltage listed in Table 44-1. Measure and record the output voltage.

Voltage-to-Current Converter

4. For each dc input voltage in Table 44-2, calculate the dc output current of Fig. 44-2, calculate the dc output current of Fig. 44-2. Record your answers.

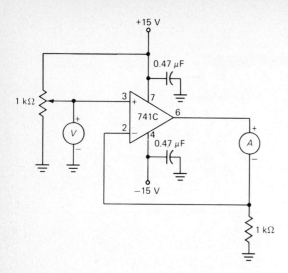

Figure 44-2

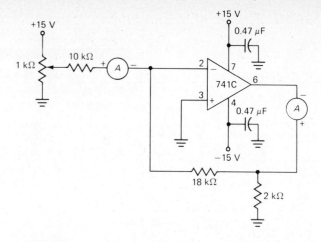

Figure 44-4

5. Connect the circuit of Fig. 44-2. Use one VOM to measure the input voltage and one VOM to measure the output current. (If you have only one VOM to work with, use a short in the place of the output ammeter when measuring the input voltage. When measuring output current, replace the short by the ammeter.)

6. Adjust the potentiometer to get an input voltage of 1 V. Read the output current and record the value in Table 44-2.

7. Repeat Step 6 for the remaining input voltages listed in Table 44-2.

Current-to-Voltage Converter

8. For each input current listed in Table 44-3, calculate the output voltage in Fig. 44-3. Record your answers.

9. Connect the circuit of Fig. 44-3.

10. Adjust the potentiometer to get an input current of 1 mA. Read the output voltage and record the value in Table 44-3.

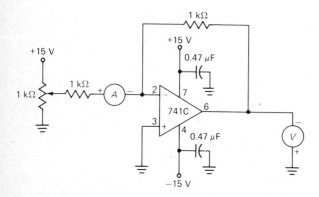

Figure 44-3

11. Repeat Step 10 for the other input currents shown in Table 44-3.

Current Amplifier

12. For each input current listed in Table 44-4, calculate the output current in Fig. 44-4. Record your answers.

13. Connect the circuit of Fig. 44-4.

14. Adjust the potentiometer to get an input current of 0.1 mA. Record the output current in Table 44-4.

15. Repeat Step 14 for the remaining input currents of Table 44-4.

Troubleshooting (Optional)

16. Ask the instructor to insert a trouble into any circuit he or she wishes.

17. Locate and repair the trouble. Record each trouble in Table 44-5.

18. Repeat Steps 16 and 17 as often as indicated by the instructor.

Design (Optional)

19. The circuit of Fig. 44-2 has a transconductance of 100 μS. Redesign the circuit so that it has a g_m of 500 μS.

20. Connect the redesigned circuit. Measure the output current for each voltage listed in Table 44-6.

Computer (Optional)

21. Write and run a program that calculates the voltage gain for Fig. 44-1, transconductance for Fig. 44-2, transresistance for Fig. 44-3, and current gain for Fig. 44-4. Your program should include a menu and whatever input statements are needed to enter necessary data.

194

DATA FOR EXPERIMENT 44

Table 44-1. Noninverting Voltage Feedback

V_{in}	Calculated V_{out}	Measured V_{out}
0.1 V		
0.2 V		
0.3 V		
0.4 V		
0.6 V		
0.8 V		
1 V		

Table 44-2. Noninverting Current Feedback

V_{in}	Calculated I_{out}	Measured I_{out}
1 V		
2 V		
3 V		
4 V		
6 V		
8 V		
10 V		

Table 44-3. Inverting Voltage Feedback

V_{in}	Calculated V_{out}	Measured V_{out}
1 mA		
2 mA		
3 mA		
4 ma		
6 mA		
8 mA		
10 mA		

Table 44-4. Inverting Current Feedback

I_{in}	Calculated I_{out}	Measured I_{out}
0.1 mA		
0.2 mA		
0.3 mA		
0.4 mA		
0.6 mA		
0.8 mA		
1 mA		

Table 44-5. Troubleshooting

Trouble	Description
1	
2	
3	

Table 44-6. Design

V_{in}	I_{out}
1 V	
2 V	
3 V	
4 V	
6 V	
8 V	
10 V	

QUESTIONS FOR EXPERIMENT 44

1. The voltage gain of Table 44-1 is closest to: (   )
   (a) 1;   (b) 5;   (c) 10;   (d) 20.
2. The transconductance of Table 44-2 is approximately: (   )
   (a) 100 μS;   (b) 300 μS;   (c) 750 μS;   (d) 1000 μS.
3. The transresistance of Table 44-3 is approximately: (   )
   (a) 100 Ω;   (b) 1 kΩ;   (c) 10 kΩ;   (d) 100 kΩ.
4. The current gain of Table 44-4 is closest to: (   )
   (a) 1;   (b) 10;   (c) 100;   (d) 1000.
5. The stability or accuracy of any of the feedback circuits in this experiment   (   )
   depends primarily on the:
   (a) supply voltage;   (b) 741C;   (c) VOM;   (d) tolerance of feedback resistors.

6. What did you learn from this experiment? List at least two ideas that seem important to you.

Troubleshooting (Optional)

7. Somebody mistakenly uses a 10-kΩ resistor for the feedback resistor of Fig. 44-2. How will this affect the circuit performance?

8. The negative supply voltage of Fig. 44-3 is not connected to the op amp. What are the symptoms of this trouble?

Design (Optional)

9. You are designing an electronic VOM. Which of the basic feedback circuits would you use to measure voltage? Which would you use for measuring current?

10. Optional. Instructor's question.

EXPERIMENT 45

GAIN-BANDWIDTH PRODUCT

▼

Whenever you work with an op amp, remember that the gain-bandwidth product is a constant. This means the product of closed-loop voltage gain and bandwidth equals the unity-gain frequency of the op amp. Stated another way, it means you can trade off voltage gain for bandwidth. For instance, if you reduce the voltage gain by a factor of 2, you will double the bandwidth.

In this experiment, you will calculate and measure the bandwidth for different voltage gains. This will confirm that the gain-bandwidth product is a constant.

REQUIRED READING

Chapter 19 (Sec. 19-6) of *Electronic Principles*, 5th ed.

EQUIPMENT

1 sine/square generator
2 power supplies: ± 15 V
6 ½-W resistors: 4.7 kΩ, 6.8 kΩ, 10 kΩ, 22 kΩ, 33 kΩ, 47 kΩ
1 op amp: 741C
2 capacitors: 0.47 μF
1 VOM (analog or digital multimeter)
1 oscilloscope
1 frequency counter

PROCEDURE

Calculating Voltage Gain and Bandwidth

1. For each R listed in Table 45-1, calculate the closed-loop voltage gain of Fig. 45-1. Record all answers.
2. The typical gain-bandwidth product (same as f_{unity}) of a 741C is 1 MHz. Calculate and record the closed-loop cutoff frequency for each R listed in Table 45-1.
3. Connect the circuit with R equal to 4.7 kΩ. Look at the output signal with an oscilloscope. With the input frequency at 100 Hz, adjust the signal level to get an output of 5 V p-p.
4. Measure the peak-to-peak input voltage. Calculate

and record A_{CL} as a measured quantity (Table 45-1).
5. Measure and record the upper cutoff frequency.
6. Repeat Steps 3 to 5 for the other values of R in Table 45-1.

MEASURING RISETIME TO GET BANDWIDTH

7. Connect the circuit of Fig. 45-1 with an R of 4.7 kΩ and a square-wave generator instead of a sine-wave generator.

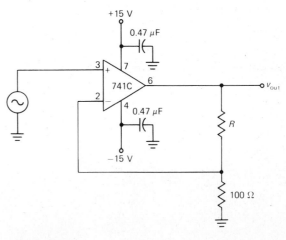

Figure 45-1

8. With the frequency around 5 kHz, adjust the signal level to get an output voltage of 5 V p-p.
9. Measure the risetime and record in Table 45-2. Calculate and record $f_{2(CL)}$.
10. Repeat Steps 7 to 9 for the other values of R. (Note: You will have to use an input frequency less than 5 kHz as the value of R increases. Reduce the frequency as needed to get an accurate risetime measurement.)

Troubleshooting (Optional)

11. Estimate the risetime in Fig. 45-1 for each trouble listed in Table 45-3. Record your answers.
12. Insert each trouble. Measure and record the risetime.

Design (Optional)

13. Select a value of R in Fig. 45-1 to get a bandwidth of 35 kHz (Use a 741C.)
14. Connect the circuit of Fig. 45-1 with your value of R. Measure the voltage gain and risetime. Calculate the bandwidth. Record all quantities listed in Table 45-4.

Computer (Optional)

15. Write and run a program that prints out the bandwidth of Fig. 45-1 for A_{CL} values that step from 10 to 100 in increments of 10 (10, 20, 30, . . . , 100). Your program should print out 10 values of bandwidth. Use an f_{unity} of 1 MHz for the 741C.

DATA FOR EXPERIMENT 45

Table 45-1. Gain and Critical Frequency

R	Calculated		Measured	
	A_{CL}	$f_{2(CL)}$	A_{CL}	$f_{2(CL)}$
4.7 kΩ				
6.8 kΩ				
10 kΩ				
22 kΩ				
33 kΩ				
47 kΩ				

Table 45-2. Risetime

R	Measured T_R	Experimental $f_{2(CL)}$
4.7 kΩ		
6.8 kΩ		
10 kΩ		
22 kΩ		
33 kΩ		
47 kΩ		

Table 45-3. Troubleshooting

Trouble	Estimated T_R	Measured T_R
R shorted		
No +15-V supply		
100 Ω open		

Table 45-4. Design

$R =$

$A_{CL} =$

$T_R =$

$f_{2(CL)} =$

QUESTIONS FOR EXPERIMENT 45

1. The measured data of Table 45-1 indicate that the product of gain and   (   )
   bandwidth is:
   (a) 1 MHz; (b) approximately constant; (c) variable; (d) none of the
   foregoing.

2. The largest value of R in Table 45-2 produces the: ( )
(**a**) smallest T_R; (**b**) largest T_R; (**c**) smallest voltage gain; (**d**) none of
the foregoing.

3. In Fig. 45-1, an increase in voltage gain leads to a: ( )
(**a**) decrease in bandwidth; (**b**) increase in bandwidth; (**c**) loss of supply
voltage; (**d**) smaller output voltage.

4. If an op amp has a higher f_{unity}, you can get more bandwidth for a given: ( )
(**a**) supply voltage; (**b**) voltage gain; (**c**) output voltage; (**d**) MPP
value.

5. To increase the bandwidth of a circuit like Fig. 45-1, you have to: ( )
(**a**) decrease the voltage gain; (**b**) increase the supply voltage; (**c**) decrease
the f_{unity}; (**d**) increase the output voltage.

6. Why is it important to know that the gain-bandwidth product is constant?

Troubleshooting (Optional)

7. Suppose one of the bypass capacitors of Fig. 45-1 shorts out. What symptoms will
you get?

8. There is no dc or ac output voltage in a circuit like Fig. 45-1. Name three possible
causes.

Design (Optional)

9. You are designing an amplifier to have as fast a risetime as possible. Do you want an
op amp with a low or a high f_{unity}? Why?

10. Optional. Instructor's question.

EXPERIMENT 46

LINEAR IC AMPLIFIERS

▼

Linear op-amp circuits preserve the shape of the input signal. If the input is sinusoidal, the output will be sinusoidal. Two basic voltage amplifiers are possible: the noninverting amplifier and the inverting amplifier. The inverting amplifier consists of a source resistance cascaded with a current-to-voltage converter. As discussed in your textbook, the closed-loop voltage gain equals the ratio of the feedback resistance to the source resistance.

In this experiment you will build and test both types of voltage amplifiers. You will also connect a noninverter/inverter with a single adjustment that allows you to vary the voltage gain.

REQUIRED READING

Chapter 20 (Secs. 20-1 to 20-3) of *Electronic Principles*, 5th ed.

EQUIPMENT

1 sine/square generator
2 power supplies ± 15 V
13 ½-W resistors: 100 Ω, two 1 kΩ, 1.1 kΩ, two 6.8 kΩ, 10 kΩ, 47 kΩ, 68 kΩ, 100 kΩ, 220 kΩ, 330 kΩ, 470 kΩ
1 potentiometer: 1 kΩ
1 op amp: 741C
4 capacitors: two 0.47 µF, two 1 µF
1 VOM (analog or digital multimeter)
1 oscilloscope
1 frequency counter

PROCEDURE

Single-Supply Noninverting Amplifier

1. Assume a typical f_{unity} of 1 MHz for the 741C of Fig. 46-1. Calculate A_{CL} and $f_{2(CL)}$. Also calculate the input, output, and bypass critical frequencies.

Estimate the MPP value. Record all answers in Table 46-1.
2. Connect the circuit. Adjust the audio generator to 100 mV p-p at 1 kHz. Measure and record A_{CL}.
3. Measure and record the upper critical frequency. (Try both the sine- and square-wave methods.)
4. Measure and record the lower critical frequency.
5. Measure and record the MPP value.

Inverting Amplifier

6. For each R value of Table 46-2, calculate A_{CL} and $f_{2(CL)}$ in Fig. 46-2.
7. Connect the circuit with R equal to 4.7 kΩ. Set the input frequency to 100 Hz. Adjust the signal level to get an 'output of 5 V p-p.
8. Measure v_{in}. Calculate the record A_{CL} as a measured quantity.
9. Measure and record $f_{2(CL)}$.
10. Repeat Steps 7 to 9 for other R values in Table 46-2.

Noninverter/Inverter

11. Calculate the maximum noninverting and inverting voltage gains for the circuit of Fig. 46-3. Record in Table 46-3.

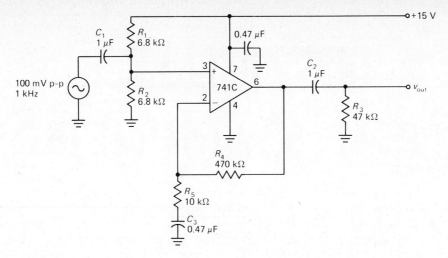

Figure 46-1

12. Connect the circuit.
13. Look at the output signal with an oscilloscope. Vary the potentiometer and notice what happens.
14. Measure the maximum noninverting and inverting voltage gains. Record the data in Table 46-3.

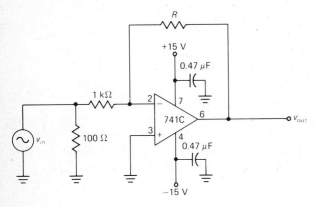

Figure 46-2

Troubleshooting (Optional)

15. For each trouble listed in Table 46-4, estimate and record the dc voltage at pin 6 (Fig. 46-1).
16. Insert each trouble into the circuit. Measure and record the dc voltage at pin 6.

Design (Optional)

17. Select new values for C_1 and C_3 to get a lower cutoff frequency in Fig. 46-1 that is less than 20 Hz.
18. Connect the circuit. Measure and record the lower cutoff frequency. Record all quantities listed in Table 46-5.

Computer (Optional)

19. Write and run a program that calculates $f_{2(CL)}$ in the inverting amplifier of Fig. 46-2 for each of these voltage gains: $A_{CL} = 1, 2, 3, \ldots, 10$.

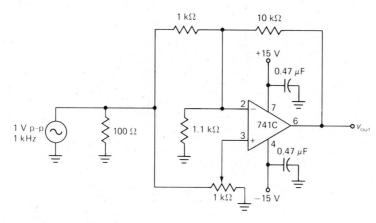

Figure 46-3

● DATA FOR EXPERIMENT 46

Table 46-1. Noninverting Amplifier

Calculated _____

$A_{CL} =$ _____

$f_{2(CL)} =$ _____

$f_{in} =$ _____

$f_{out} =$ _____

$f_{BY} =$ _____

$MPP =$ _____

Measured _____

$A_{CL} =$ _____

$F_{2(CL)} =$ _____

$F_{1(CL)} =$ _____

$MPP =$ _____

● Table 46-2

R	Calculated		Measured	
	A_{CL}	$f_{2(CL)}$	A_{CL}	$f_{2(CL)}$
47 kΩ				
68 kΩ				
100 kΩ				
220 kΩ				
330 kΩ				
470 kΩ				

Table 46-3. Noninverter/Inverter

Calculated _____

$A_{non} =$ _____

$A_{inv} =$ _____

Measured _____

$A_{non} =$ _____

● $A_{inv} =$ _____

Table 46-4. Troubleshooting

Trouble	DC Voltage at Pin 6
R_1 open	
R_1 short	
R_2 open	
R_2 short	
C_1 open	

Table 46-5. Design

$C_1 =$	
$C_3 =$	
$f_{1(CL)} =$	

QUESTIONS FOR EXPERIMENT 46

1. The MPP value in Table 46-1 is closest to: (   )
   (a) 1 V;   (b) 7.5 V;   (c) 12.5 V;   (d) 20 V.
2. The product of voltage gain and bandwidth in Table 46-2 is: (   )
   (a) approximately constant;   (b) small;   (c) 100;   (d) 20 kHz.
3. The noninverter/inverter of Fig. 46-3 has a noninverting voltage gain of   (   )
   approximately:
   (a) 1;   (b) 10;   (c) 100;   (d) 1000.
4. The bypass capacitor of Fig. 46-1 sets up a cutoff frequency of approximately: (   )
   (a) 3.39 Hz;   (b) 33.9 Hz;   (c) 46.8 Hz;   (d) 63 Hz.
5. With the inverter of Fig. 46-2, you can increase the bandwidth by: (   )
   (a) decreasing the supply voltage;   (b) decreasing the voltage gain;
   (c) increasing the value of R;   (d) eliminating the bypass capacitors.
6. Explain how the inverter of Fig. 46-2 works.

Troubleshooting (Optional)

7. What happens to the dc voltage at pin 6 when R_1 is shorted? Why?

8. Suppose the bypass capacitor C_3 opens in Fig. 46-1. What kind of dc and ac symptoms will you get at the output of the op amp?

Design (Optional)

9. Explain why you selected the C_1 and C_3 in your design.

10. Optional. Instructor's question.

EXPERIMENT 47

CURRENT SOURCE AND ACTIVE FILTER

With op amps we can build accurate voltage-controlled current sources. Your textbook shows several designs that can provide either unilateral or bilateral current to a load. In this experiment you will connect the grounded voltage-to-current converter of Sec. 20-6.

To avoid large inductors, op amps can be used to build active filters with low cutoff frequencies. Depending on the number of poles, an active filter can provide a fast rolloff in response. In this experiment, you will connect a two-pole Butterworth low-pass filter.

REQUIRED READING

Chapter 20 (Secs. 20-6 and 20-8) of *Electronic Principles*, 5th ed.

EQUIPMENT

- 1 audio generator
- 2 power supplies ± 15 V
- 11 ½-W resistors: 100 Ω, three 1 kΩ, 1.2 kΩ, 1.8 kΩ, 2.2 kΩ, 4.7 kΩ, 10 kΩ, two 33 kΩ
- 1 potentiometer: 1 kΩ
- 1 op amp: 741C
- 4 capacitors: two 1000 pF, two 0.47 μF
- 1 VOM (analog or digital multimeter)
- 1 oscilloscope
- 1 frequency counter

PROCEDURE

Voltage-Controlled Current Source

1. In Fig. 47-1, assume R_L is 100 Ω and calculate the load current for each value of input voltage listed in Table 47-1. Record your answers.
2. Assume R_1 is 2.2 kΩ and calculate the load current for each value of input voltage shown in Table 47-2. Record your answers.
3. Connect the circuit with R_L equal to 100 Ω.

4. Adjust the input voltage to each value given in Table 47-1. Measure the load current. Record your data.
5. Change R_L to 2.2 kΩ. Repeat Step 4 using Table 47-2.

Two-Pole Low-Pass Butterworth Filter

6. Calculate the cutoff frequency of Fig. 47-2. Record at the top of Table 47-3.
7. Connect the circuit.
8. Adjust the input signal level to 1 V p-p.
9. Look at the output with an oscilloscope. As you increase the frequency from 0 to 20 kHz, you should see the response drop off.
10. Set the frequency to 100 Hz. Adjust the signal level to get an output of 1 V rms (use the VOM as an ac voltmeter).
11. Measure and record the output voltage for each frequency listed in Table 47-3. (Keep the input voltage the same as it is for 100 Hz.)
12. If possible, increase the signal frequency to 50 kHz. Measure the output and notice that it is down approximately 40 dB from the low-frequency value.

Troubleshooting (Optional)

13. In Fig. 47-1 assume an R_L of 100 Ω and a v_{in} of + 5 V. Estimate and record the load current for each trouble listed in Table 47-4.
14. Insert each trouble into the circuit. Measure and record the load current.

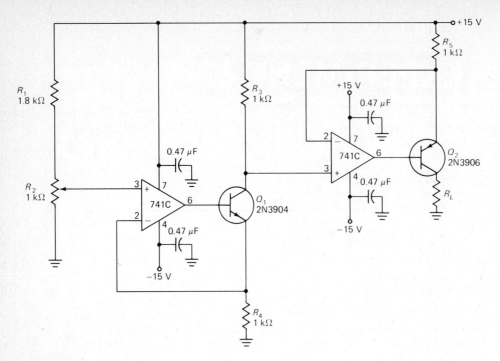

Figure 47-1

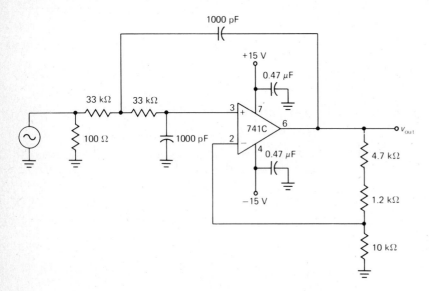

Figure 47-2

Design (Optional)

15. Redesign the circuit of Fig. 47-1 to get a load current of approximately 2.25 mA when v_{in} is +5 V.

16. Connect the circuit with your design values. Measure the load current with a v_{in} of +5 V. Record all quantities listed in Table 47-5.

Computer (Optional)

17. Write and run a program that calculates the load current in Fig. 47-1.

DATA FOR EXPERIMENT 47

Table 47-1. Voltage-Controlled Current Source:
$R_L = 100\ \Omega$

V_{in}	Calculated I_L	Measured I_L
0 V		
1 V		
2 V		
3 V		
4 V		
5 V		

Table 47-2. Voltage-Controlled Current Source:
$R_L = 2.2\ k\Omega$

V_{in}	Calculated I_L	Measured I_L
0 V		
1 V		
2 V		
3 V		
4 V		
5 V		

Table 47-3. Two-Pole Low-Pass Butterworth Filter:

$f_c =$ _____

f	v_{out}
100 Hz	1 V
1 kHz	
2 kHz	
3 kHz	
4 kHz	
6 kHz	
8 kHz	
10 kHz	
15 kHz	
20 kHz	

Table 47-4. Troubleshooting

Trouble	Estimated I_L	Measured I_L
R_3 short		
R_4 open		
Q_1 open		
Q_2 open		

Table 47-5. Design

$R =$

$I_L =$

QUESTIONS FOR EXPERIMENT 47

1. The measured data of Tables 47-1 and 47-2 indicate that the load resistance ( )
   is driven by a:
   (a) voltage source; (b) current source; (c) transistor; (d) op amp.
2. The last measured entry of Table 47-2 indicates that the load voltage exceeds: ( )
   (a) V_{CC}; (b) $V_{CC} - v_{in}$; (c) V_{EE}; (d) $I_{out(max)}$.
3. As R_L increases in Fig. 47-1, the maximum input voltage: ( )
   (a) decreases; (b) increases; (c) stays the same; (d) equals zero.
4. The low-pass filter response of Fig. 47-2 rolls off at a rate of: ( )
   (a) 6 dB per octave; (b) 12 dB per decade; (c) 20 dB per decade;
   (d) 40 dB per decade.
5. The cutoff frequency of Fig. 47-2 is closest to: ( )
   (a) 100 Hz; (b) 1 kHz; (c) 5 kHz; (d) 20 kHz.
6. Explain why the circuit of Fig. 47-1 cannot produce 5 mA when R_L is 2.2 kΩ.

Troubleshooting (Optional)

7. Why does the load current decrease to zero when R_3 is shorted in Fig. 47-1?

8. Somebody has connected the circuit of Fig. 47-2 without the 100-Ω resistor across
   the source. The circuit may or may not work. If the circuit does not work, what do
   you think the trouble is?

Design (Optional)

9. If you want to double the cutoff frequency of Fig. 47-2, what changes do you need to make?

10. Optional. Instructor's question.

▲ ▲

EXPERIMENT 48

ACTIVE DIODE CIRCUITS AND COMPARATORS

▼

With op amps we can reduce the effect of diode knee voltage. The effective knee voltage is reduced by the open-loop gain of the op amp. For a typical 741C, this means the equivalent knee voltage is only 7 μV. This allows us to build circuits that will rectify, peak-detect, limit, and clamp low-level signals.

A comparator is a circuit that can indicate when the input voltage exceeds a specific limit. With a zero-crossing detector, the trip point is zero. With a limit detector, the trip point is either a positive or negative voltage.

In this experiment you will build a variety of active diode circuits as well as a zero-crossing detector and a limit detector.

▲ ▲

REQUIRED READING

Chapter 21 (Secs. 21-1 and 21-2) of *Electronic Principles*, 5th ed.

EQUIPMENT

1 audio generator
2 power supplies: ±15 V
6 ½-W resistors: 100 Ω, 1 kΩ, 2.2 kΩ, two 10 kΩ, 100 kΩ
1 potentiometer: 1 kΩ
1 diode: 1N914
2 LEDs: TIL221 and TIL222 (or similar red and green LEDs)
1 op amp: 741C
3 capacitors: two 0.47 μF, 100 μF (rated at least 15 V)
1 VOM (analog or digital multimeter)
1 oscilloscope

PROCEDURE

Half-Wave Rectifier

1. Build the circuit of Fig. 48-1.
2. Connect the oscilloscope (dc input) across the

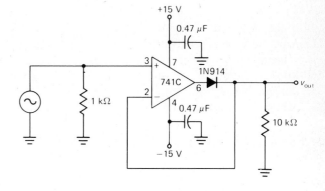

Figure 48-1

10-kΩ load resistor. Set the generator to 100 Hz and adjust the level to get a peak output of 1 V on the oscilloscope. (You should be looking at a half-wave signal.)

3. Measure the peak value of the input sine wave. Record the input and output peak voltages in Table 48-1.
4. Adjust the signal level to get a half-wave output with a peak value of 100 mV. Then measure the peak input voltage. Record the input and output peak voltages in Table 48-1.
5. Reverse the polarity of the diode. The output voltage should be a negative half-wave signal.

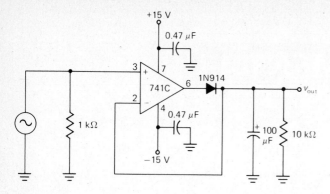

Figure 48-2

Peak Detector

6. Connect a 100-μF capacitor across the load to get the circuit of Fig. 48-2.

7. Adjust the generator to get an input peak value of 1 V. Measure the dc output voltage. Record the peak input voltage and the dc output voltage in Table 48-2.

8. Readjust the generator to get an input peak value of 100 mV. Measure the dc output voltage. Record the peak input voltage and the dc output voltage in Table 48-2.

9. Reverse the polarity of the diode and the capacitor. You should get a negative dc output voltage.

Limiter

10. Build the circuit of Fig. 48-3.

11. Adjust the generator to produce a peak-to-peak value of 1 V at the left end of the 2.2-kΩ resistor.

12. Look at the output signal while turning the potentiometer through its entire range.

13. Adjust the generator to produce a peak-to-peak output of 100 mV at the left end of the 2.2-kΩ resistor. Then repeat Step 12.

14. Reverse the polarity of the diode and repeat Step 12 for a peak-to-peak output of 1 V.

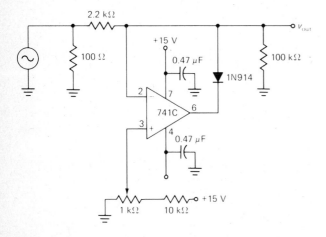

Figure 48-3

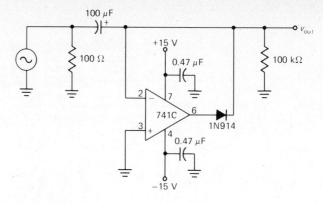

Figure 48-4

DC Clamper

15. Connect the circuit of Fig. 48-4.

16. Adjust the input to 1 V p-p.

17. Look at the output. It should be a positively clamped signal.

18. Reduce the input to 100 mV p-p and repeat Step 17.

19. Reverse the polarity of the diode. The output should now be negative clamped.

Zero-Crossing Detector

20. Connect the zero-crossing detector of Fig. 48-5. (Note: The I_{max} out of the 741C is approximately 25 mA, so the LED current is limited to 25 mA. If an op amp has an I_{max} greater than 50 mA, you would need current-limiting resistors because most LEDs cannot handle more than 50 mA.)

21. Vary the potentiometer and notice what the LEDs do.

22. Use the dc-coupled input of the oscilloscope to look at the input voltage to pin 3. Adjust the potentiometer to get +100 mV at the input. Record the input voltage and the color of the on LED (Table 48-3).

23. Adjust the potentiometer to get an input of −100 mV. Record the input voltage and the color of the on LED.

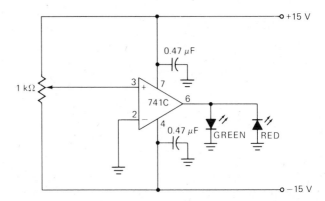

Figure 48-5

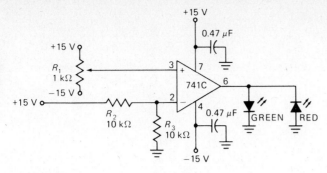

Figure 48-6

Limit Detector

24. In Fig. 48-6, calculate the trip point of the limit detector. Record your answer in Table 48-4.

25. Connect the circuit. Adjust the input voltage until you locate the approximate trip point. Record the trip point.

Troubleshooting (Optional)

26. For each set of symptoms listed in Table 48-5, try to figure out what trouble could produce the symptoms in Fig. 48-6. Insert the trouble and verify that it produces the symptoms. Record each trouble in Table 48-5.

Design (Optional)

27. Select a value of R_3 to get a trip point of approximately +5 V. Connect the circuit with your design value and measure the trip point. Record the quantities listed in Table 48-6.

Computer (Optional)

28. Write and run a program that calculates the trip point in Fig. 48-6 for different values of R_2 and R_3.

DATA FOR EXPERIMENT 48

Table 48-1. Active Half-Wave Rectifier

Step 3:

v_{in} = _____

v_{out} = _____

Step 4:

v_{in} = _____

v_{out} = _____

Table 48-2. Active Peak Detector

Step 7:

v_{in} = _____

v_{out} = _____

Step 8:

v_{in} = _____

v_{out} = _____

Table 48-3. Zero-Crossing Detector

Step 22:

v_{in} = _____

Color = _____

Step 23:

v_{in} = _____

Color = _____

Table 48-4. Limit Detector

Calculated trip point = _____

Measured trip point = _____

Table 48-5. Troubleshooting

Symptoms	Trouble
1. Red LED always on	
2. Trip point equals zero	
3. Red LED goes on and off, but green LED is always off	
4. Neither LED comes on	

Table 48-6. Design

$R_3 =$

Trip point =

QUESTIONS FOR EXPERIMENT 48

1. The circuit of Fig. 48-1 is a: ( )
   (a) half-wave rectifier; (b) full-wave rectifier; (c) bridge rectifier;
   (d) none of the foregoing.
2. The dc output voltage of Fig. 48-2 is approximately equal to the: ( )
   (a) peak input voltage; (b) positive supply voltage; (c) rms input voltage; (d) average input voltage.
3. The positive limiter of Fig. 48-3 can be adjusted to have a limiting level ( )
   between 0 and approximately:
   (a) 0; (b) +1.36 V; (c) −5 V; (d) +12 V.
4. The circuit of Fig. 48-4 clamps the signal: ( )
   (a) negatively; (b) positively; (c) at −5 V; (d) at +3 V.
5. The limit detector of Fig. 48-6 has a trip point of approximately: ( )
   (a) 0; (b) +5 V; (c) +7.5 V; (d) +10 V.
6. Explain how the limit detector of Fig. 48-6 works.

Troubleshooting (Optional)

7. Name at least two troubles in Fig. 48-6 that would produce a trip point of zero.

8. If the capacitor of Fig. 48-2 opens, what symptoms will you get?

Design (Optional)

9. How did you arrive at your design value for R_3?

10. Optional. Instructor's question.

220

WAVESHAPING CIRCUITS

▼

By using positive feedback with a comparator, we can build a Schmitt trigger. It has hysteresis, which makes it less sensitive to noise. A Schmitt trigger is useful for waveshaping because it produces a square-wave output no matter what the shape of the input signal.

If we add an *RC* circuit to a Schmitt trigger, we get a relaxation oscillator. This type of circuit generates a square-wave output without an external input signal. By cascading a relaxation oscillator and an integrator, we can build a circuit that generates square waves and triangular waves.

REQUIRED READING

Chapter 21 (Secs. 21-3 to 21-6) of *Electronic Principles*, 5th ed.

EQUIPMENT

1 audio generator
2 power supplies: adjustable from 0 to 15 V
8 ½-W resistors: 100 Ω, 1 kΩ, two 2.2 kΩ, 10 kΩ, 18 kΩ, 22 kΩ, 100 kΩ
3 op amp: 318C, two 741C
6 capacitors: two 0.1 μF, four 0.47 μF
1 oscilloscope
1 frequency counter

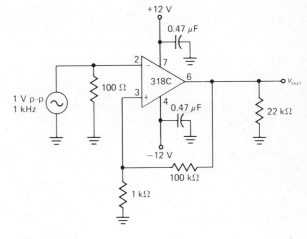

Figure 49-1

PROCEDURE

Schmitt Trigger

1. In Fig. 49-1, what shape do you think the output signal will have? Estimate the peak-to-peak output voltage. Record your answers in Table 49-1. Also calculate and record the trip points.
2. Connect the circuit. Adjust the input voltage to 1 V p-p at 1 kHz.
3. Look at the output with an oscilloscope. Record the approximate shape of the signal in Table 49-1. Also measure and record the peak-to-peak output voltage.
4. Look at the noninverting input voltage with the

oscilloscope on dc input. Measure the positive peak and record this as the UTP. Measure the negative peak and record as the LTP.

Effect of Slew-Rate Limiting

5. Increase the frequency to 20 kHz. The output should be approximately rectangular. (Note: You may see some overshoot or ringing because of the high slew rate of a 318C, but the transitions between high and low should still appear almost vertical.)
6. Change the frequency back to 1 kHz. Replace the 318C by a 741C. The output should appear approximately rectangular.

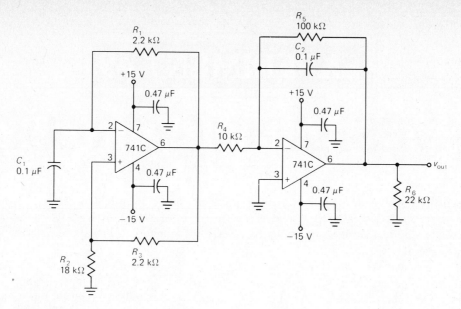

Figure 49-2

7.
Increase the frequency and notice how the slew rate of a 741C affects the vertical transitions.

Relaxation Oscillator and Integrator

8. In Fig. 49-2, a relaxation oscillator drives an integrator. Calculate the frequency out of the relaxation oscillator. Record in Table 49-2.

9. Assume $+V_{SAT}$ is $+14$ V and $-V_{SAT}$ is -14 V. Calculate and record the peak-to-peak output voltage from the relaxation oscillator.

10. Calculate and record the peak-to-peak output voltage from the integrator.

11. Connect the circuit.

12. Look at the signal out of the relaxation oscillator. Measure its frequency and peak-to-peak value. Record these data.

13. Look at the signal at the inverting input of the relaxation oscillator. It should look like Fig. 21-23b in your textbook.

14. Look at the signal out of the integrator. Measure and record its peak-to-peak value.

Troubleshooting (Optional)

15. For each trouble listed in Table 49-3, calculate the output frequency and the peak-to-peak output voltage in Fig. 49-2. Record your answers.

16. Insert each trouble into the circuit. Measure and record the output frequency and peak-to-peak voltage.

Design (Optional)

17. Select a value of R_1 in Fig. 49-2 to get a frequency of approximately 1 kHz.

18. Connect the circuit with your design value of R_1. Measure the frequency. Record your value of R_1 and frequency in Table 49-4.

Computer (Optional)

19. Write and run a program for a circuit like Fig. 49-2 that calculates the frequency and peak-to-peak value of the triangular output. Use INPUT statements to enter whatever data are required.

DATA FOR EXPERIMENT 49

Table 49-1. Schmitt Trigger

Calculated:

Shape = _____

MPP = _____

UTP = _____

LTP = _____

Measured:

Shape = _____

MPP = _____

UTP = _____

LTP = _____

Table 49-2. Relaxation Oscillator and Integrator

Calculated

f = _____

$v_{out(1)}$ = _____

$v_{out(2)}$ = _____

Measured:

f = _____

$v_{out(1)}$ = _____

$v_{out(2)}$ = _____

Table 49-3. Troubleshooting

Trouble	Calculated		Measured	
	f	v_{out}	f	v_{out}
R_1 is 22 kΩ				
R_2 is 1.8 kΩ				
R_4 is 22 kΩ				

Table 49-4. Design

R_1 =

f =

QUESTIONS FOR EXPERIMENT 49

1. The square wave out of the Schmitt trigger (Fig. 49-1) has a peak-to-peak ( )
   value that is closest to:
   (a) 5 V; (b) 10 V; (c) 20 V; (d) 30 V.
2. The UTP of the Schmitt trigger was approximately: ( )
   (a) −0.1 V; (b) +0.1 V; (c) −10 V; (d) +10 V.
3. The relaxation oscillator has a calculated frequency in Table 49-2 of: ( )
   (a) 345 Hz; (b) 456 Hz; (c) 796 Hz; (d) 1.27 kHz.
4. The triangular output of the integrator in Fig. 49-2 has a calculated peak-to- ( )
   peak value of approximately:
   (a) 0.1 V; (b) 8.79 V; (c) 12.3 V; (d) 15 V.
5. The waveform at the inverting input of the relaxation oscillator (Fig. 49-2) ( )
   appears:
   (a) square; (b) triangular; (c) exponential; (d) sinusoidal.
6. Explain how a Schmitt trigger like Fig. 49-1 works.

7. Explain how a relaxation oscillator like Fig. 49-2 works.

Troubleshooting (Optional)

8. What are the symptoms in Fig. 49-2 when R_1 is 22 kΩ instead of 2.2 kΩ? Why do
   these changes occur?

224

Design (Optional)

9. Why is it better to use a 318C instead of a 741C in a Schmitt trigger?

10. Optional. Instructor's question.

EXPERIMENT 50

THE WIEN-BRIDGE OSCILLATOR

▼

The Wien-bridge oscillator is the standard oscillator circuit for low to moderate frequencies in the range of 5 Hz to about 1 MHz. The oscillation frequency is equal to $1/2\pi RC$. Typically, a tungsten lamp is used to reduce the loop gain AB to unity. It is also possible to use diodes, zener diodes, and JFETs as nonlinear elements to reduce the loop gain to unity. In this experiment you will build and test a Wien-bridge oscillator.

REQUIRED READING

Chapter 22 (Secs. 22-1 and 22-2) of *Electronic Principles*, 5th ed.

EQUIPMENT

2 power supplies: ±15 V
9 ½-W resistors: two 1 kΩ, two 2.2 kΩ, two 4.7 kΩ, 8.2 kΩ, two 10 kΩ
3 diodes: 1N914
1 LED: TIL221 (or similar red LED)
1 op amp: 741C
1 potentiometer: 5 kΩ

4 capacitors: two 0.01 μF, two 0.47 μF
1 oscilloscope
1 frequency counter

PROCEDURE

Oscillator

1. In Fig. 50-1, calculate and record the oscillation for each value of R listed in Table 50-1. Also calculate and record the MPP value. (Note: The LED is used to indicate when the circuit is oscillating. The 1N914 across the LED protects it during reverse bias because the LED has a breakdown voltage of only 3 V.)

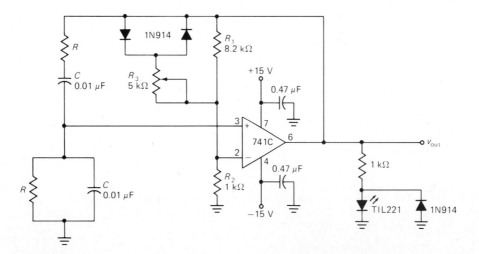

Figure 50-1

2. Connect the circuit with an R of 10 kΩ. Look at the output with an oscilloscope. Adjust R_3 to get as large an unclipped output as possible.
3. Measure the frequency. Measure the peak-to-peak output voltage. Record both quantities in Table 50-1.
4. Repeat Steps 2 and 3 for the other values of R.

Troubleshooting (Optional)

5. Insert each trouble listed in Table 50-2. Determine what effect the trouble has on the output signal. Record the symptoms in Table 50-2. (Examples of entries are "no output," "heavily clipped output," "small distorted output," etc.)

Design (Optional)

6. Select a value of R (nearest standard value) to get an oscillation frequency of approximately 2.25 kHz.
7. Connect the circuit with your value of R. Measure the frequency. Record the quantities of Table 50-3.

Computer (Optional)

8. Write and run a program that calculates the oscillation frequency of a Wien-bridge oscillator.

DATA FOR EXPERIMENT 50

Table 50-1. Oscillator

	Calculated			Measured	
R	f	MPP		f	MPP
10 kΩ					
4.7 kΩ					
2.2 kΩ					

Table 50-2. Troubleshooting

Trouble	Symptoms
R_1 short	
R_1 open	
R_2 short	
R_2 open	
R_3 short	
R_3 open	

Table 50-3. Design

$R =$

$f =$

QUESTIONS FOR EXPERIMENT 50

1. The data of Table 50-1 indicate that an increase in resistance produces which   (   )
   of the following changes in oscillation frequency:
   (**a**) decrease;   (**b**) increase;   (**c**) no change.
2. The MPP value of the circuit was closest to: (   )
   (**a**) 0.7 V;   (**b**) 1.4 V;   (**c**) 15 V;   (**d**) 27 V.
3. Open component that is involved in reducing loop gain to unit is the: (   )
   (**a**) 741C;   (**b**) LED;   (**c**) 1N914;   (**d**) 0.01 μF.
4. The peak LED current is closest to: (   )
   (**a**) 8.59 mA;   (**b**) 17.1 mA;   (**c**) 19.1 mA;   (**d**) 27 mA.
5. The 1N914 protects the LED from excessive reverse voltage because the   (   )
   1N914:
   (**a**) breaks down first;   (**b**) conducts when the reverse voltage exceeds
   -0.7 V;   (**c**) goes into reverse bias when the LED is on;   (**d**) has a higher
   power dissipation than the LED.

6. Briefly explain how a Wien-bridge oscillator works.

Troubleshooting (Optional)

7. Explain why there is no output when R_1 is shorted.

8. Explain why the output is heavily clipped when R_3 is open.

Design (Optional)

9. How can you make the Wien-bridge oscillator tunable to different frequencies?

10. Optional. Instructor's question.

EXPERIMENT 51

THE *LC* OSCILLATOR

▼

For oscillation frequencies between approximately 1 and 500 MHz, the *LC* oscillator is used instead of a Wien-bridge oscillator. This type of oscillator uses a resonant *LC* tank circuit to determine the frequency. The Colpitts oscillator is the most widely used *LC* oscillator because the feedback voltage is conveniently produced by a capacitive voltage divider rather than an inductive divider (Hartley). For an oscillator to start, the small-signal voltage gain must be greater than the reciprocal of the feedback fraction. In symbols, $A > 1/B$. As the oscillations increase, the value of A decreases until the loop gain AB equals unity.

REQUIRED READING

Chapter 22 (Sec. 22-4) of *Electronic Principles*, 5th ed.

EQUIPMENT

1 power supply: 15 V
4 ½-W resistors: 4.7 kΩ, 10 kΩ, 22 kΩ, 47 kΩ
1 inductor: 100 μH
4 capacitors: 0.001 μF, 0.01 μF, 0.1 μF, 0.47 μF
1 transistor: 2N3904
1 oscilloscope
1 frequency counter

PROCEDURE

Colpitts Oscillator

1. In Fig. 51-1, neglect transistor and stray-wiring capacitance. Calculate the frequency of oscillation. Also calculate the peak-to-peak output voltage and the feedback fraction. Record your answers in Table 51-1. (Note: The 0.47-μF capacitor is a supply bypass capacitor needed with some power supplies. This capacitor ac grounds the upper end of the 100 μH and prevents the supply impedance from affecting the oscillation frequency and amplitude.)
2. Assume a load resistance of 10 kΩ is across the output. Calculate and record the ac load power.
3. Connect the circuit.

4. Look at the output signal with an oscilloscope. You should see a large sinusoidal signal.
5. Measure and record the frequency of oscillation. Measure and record the peak-to-peak output voltage.
6. Look at the signal on emitter. Measure the peak-to-peak value of this signal. Then calculate the feedback fraction. Record this as the measured B in Table 51-1.
7. Connect a load resistance of 10 kΩ across the output. Measure the peak-to-peak output voltage. Calculate

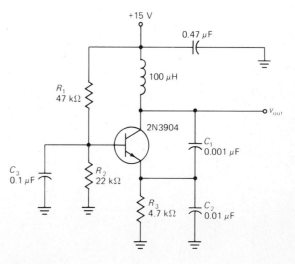

Figure 51-1

231

the ac load power and record this as the measured load power.

Troubleshooting (Optional)

8. Remove the 10-kΩ load resistance. Insert each trouble listed in Table 51-2. Record the output symptoms. Examples of entries are "no output," "smaller output," "higher frequency," etc.

Design (Optional)

9. Ignore transistor and stray capacitance. Select new values for C_1 and C_2 to get an oscillation frequency of approximately 1 MHz.

10. Insert your design values. Measure the oscillation frequency. Record all quantities in Table 51-3.

Computer (Optional)

11. Write and run a program for Fig. 51-1 that calculates the frequency of oscillation, the feedback fraction, and the minimum voltage gain needed for oscillations to start.

DATA FOR EXPERIMENT 51

Table 51-1. Colpitts Oscillator

Calculated	Measured
$f =$	$f =$
MPP =	MPP =
$B =$	$B =$
$P_{load} =$	$P_{load} =$

Table 51-2. Troubleshooting

Trouble	Output Symptoms
R_1 short	
R_1 open	
R_2 short	
R_2 Open	
R_3 short	
R_3 open	
C_1 short	
C_1 open	
C_2 short	
C_2 open	
C_3 short	
C_3 open	

Table 51-3. Design

$C_1 =$	
$C_2 =$	
$f =$	

QUESTIONS FOR EXPERIMENT 51

1. The calculated oscillation frequency of Fig. 51-1 is approximately: ( )
   (a) 100 kHz; (b) 225 kHz; (c) 445 kHz; (d) 528 kHz.
2. The calculated feedback fraction of Fig. 51-1 is closest to: ( )
   (a) 0.091; (b) 0.1; (c) 1; (d) 10.
3. For the oscillator to start, the minimum voltage gain is approximately: ( )
   (a) 1; (b) 5; (c) 11; (d) 25.

4. The LC oscillator of Fig. 51-1 is an example of a:　　　　　( )
 (**a**) CB oscillator;　(**b**) CE oscillator;　(**c**) Wien bridge;　(**d**) twin T.

5. The calculated peak-to-peak output voltage of Fig. 51-1 is approximately:　　( )
 (**a**) 20 V;　(**b**) 25 V;　(**c**) 30 V;　(**d**) 40 V.

6. Briefly explain how an LC oscillator works.

7. In Fig. 51-1, what effect will transistor and stray-wiring capacitance have on the frequency of oscillation?

Troubleshooting (Optional)

8. Why is there no output when R_1 is open?

Design (Optional)

9. Explain why the actual frequency of oscillation will be less than your calculated frequency of oscillation.

10. Optional. Instructor's question.

EXPERIMENT 52

UNWANTED OSCILLATIONS

▼

When you build a high-gain three-stage amplifier, you are likely to get unwanted oscillations unless special precautions are taken. As described in your textbook, you may get motorboating (low-frequency oscillations) because of power-supply impedance or you may get high-frequency oscillations caused by lead inductance, interstage coupling, or ground loops. In this experiment, you will build a three-stage amplifier that produces unwanted oscillations.

REQUIRED READING

Chapter 22 (Sec. 22-7) of *Electronic Principles*, 5th ed.

EQUIPMENT

1 power supply: 15 V
3 transistors: 2N3904
7 ½-W resistors: 10 Ω, three 1 kΩ, three 220 kΩ
3 capacitors: 1 μF, two 47 μF (25-V rating or better)
1 oscilloscope
4 feet of hookup wire

PROCEDURE

1. Build the three-stage amplifier of Fig. 52-1.
2. Use the oscilloscope to look at the output of the amplifier. You should be getting motorboating because the 10-Ω resistor simulates an excessively high power-supply impedance. Measure and record the period of this motorboating (Table 52-1).

3. Replace the 10-Ω resistor by 4 ft of hookup wire. This long lead produces excessive lead inductance as described in your textbook.
4. You now should be getting high-frequency oscillations. Measure and record the period (Table 52-1).
5. Connect a 1-μF capacitor across the supply input to the amplifier. (Refer to Fig. 22-24 in your textbook.) This may or may not eliminate the unwanted oscillations.
6. Remove the 4 ft of hookup wire and the 1-μF capacitor. Reconnect the power supply with as short a lead as possible. If oscillations are still present, try connecting the 1-μF capacitor across the supply input to the amplifier. Again, this may or may not eliminate the oscillations. If not, the oscillations may be produced by ground loops or by interstage coupling.
7. Optional: If instructor desires, try to eliminate the oscillations (if they are still present) by using a single-point ground system, by increasing the distance between stages, by shielding, etc.

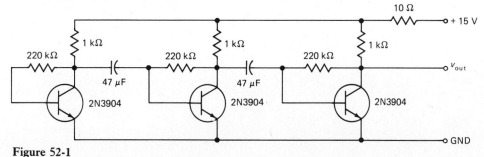

Figure 52-1

235

DATA FOR EXPERIMENT 52

Table 52-1. Period of Oscillations

Step 2 _____

Step 4 _____

QUESTIONS FOR EXPERIMENT 52

1. Motorboating is caused by: (   )
   (a) supply impedance;   (b) interstage coupling;   (c) long supply lead;
   (d) ground loops.
2. The 10-Ω resistor simulates: (   )
   (a) power-supply impedance;   (b) interstage coupling;   (c) ground loops;
   (d) long supply lead.
3. The 4 ft of hookup wire produced: (   )
   (a) ground loops;   (b) excessive lead inductance;   (c) interstage coupling;   (d) low supply voltage.
4. The 1-μF capacitor eliminates high-frequency oscillations caused by: (   )
   (a) excessive supply lead length;   (b) ground loops;   (c) interstage coupling;   (d) supply resistance.
5. Ground loops produce: (   )
   (a) low-frequency oscillations;   (b) high-frequency oscillations;   (c) motorboating;   (d) supply resistance.
6. If the first two stages each have a voltage gain of 120 and the last stage has a   (   )
   voltage gain of 200, the overall gain is:
   (a) 14,400;   (b) 256,000;   (c) 2,880,000;   (d) 5,000,000.
7. Because the voltage gain of the three-stage amplifier is so high, it is easy to: (   )
   (a) avoid oscillations;   (b) get oscillations;   (c) build nonoscillating high-
   gain amplifiers;   (d) prevent oscillations by not using an input signal.
8. Which of the following cannot stop high-frequency oscillations?   (   )
   (a) shielding;   (b) increasing the distance between stages;   (c) using a single-point ground system;   (d) using a regulated power supply.
9. Optional. Instructor's question.

10. Optional. Instructor's question.

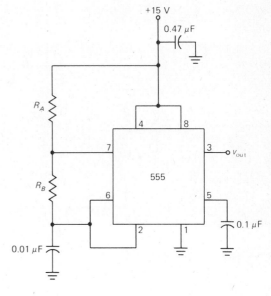

EXPERIMENT 53

THE 555 TIMER

▼

The 555 timer combines a relaxation oscillator, two comparators, and an *RS* flip-flop. This versatile chip can be used as an astable multivibrator, monostable multivibrator, VCO, ramp generator, etc. In this experiment you will build and test some basic 555 timer circuits.

REQUIRED READING

Chapter 22 (Sec. 22-9) of *Electronic Principles*, 5th ed.

EQUIPMENT

- 1 audio generator
- 1 power supply: 15 V
- 10 ½-W resistors: two 1 kΩ, 4.7 kΩ, two 10 kΩ, 22 kΩ, 33 kΩ, 47 kΩ, 68 kΩ, 100 kΩ
- 1 potentiometer: 1 kΩ
- 4 capacitors: 0.01 μF, 0.1, μF, two 0.47 μF
- 1 transistor: 2N3906
- 1 op amp: 741C
- 1 timer: NE555
- 1 oscilloscope
- 1 frequency counter

PROCEDURE

Astable 555 Timer

1. Calculate the frequency and duty cycle in Fig. 53-1 for the resistances listed in Table 53-1. Record your answers.
2. Connect the circuit of Fig. 53-1 with $R_A = 10$ kΩ and $R_B = 100$ kΩ.
3. Look at the output with an oscilloscope. Measure and record the frequency.
4. Measure *W*. Calculate and record the duty cycle as the measured *D* in Table 53-1.
5. Look at the voltage across the timing capacitor (pin 6). You should see an exponentially rising and falling wave between 5 and 10 V.
6. Repeat Steps 2 through 5 for the other resistances of Table 53-1.

Figure 53-1

Voltage-Controlled Oscillator

7. Connect the VCO of Fig. 53-2.
8. Look at the output with an oscilloscope.
9. Vary the 1-kΩ potentiometer and notice what happens. Measure and record the minimum and maximum frequencies in Table 53-2.

Monostable 555 Timer

10. Figure 53-3 shows a Schmitt trigger driving a monostable 555 timer. Assume it produces a normal trigger input for the 555. Calculate and record the pulse width out of the 555 timer for each *R* listed in Table 53-3.

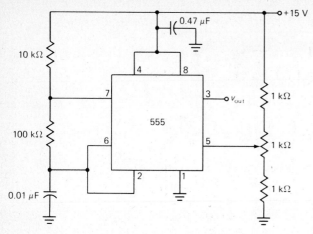

Figure 53-2

11. Connect the circuit of Fig. 53-3 with an *R* of 33 kΩ.
12. Look at the output of the Schmitt trigger (pin of 6 of the 741C). Set the frequency of the sine-wave input to 1 kHz. Adjust the sine-wave level until you get a Schmitt-trigger output with a duty cycle of approximately 90 percent.
13. Look at the output of the 555 timer. Measure and record the pulse width.
14. Repeat Steps 11 through 13 for the remaining *R* values of Table 53-3.

Ramp Generator

15. Figure 53-4 shows a ramp generator. As before, the Schmitt trigger drives a 555 timer connected for

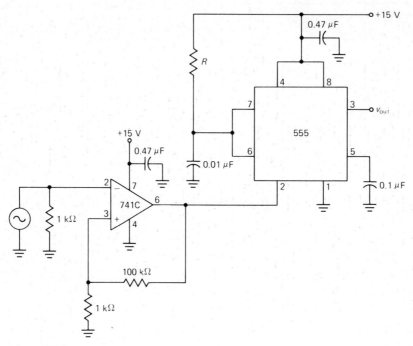

Figure 53-3

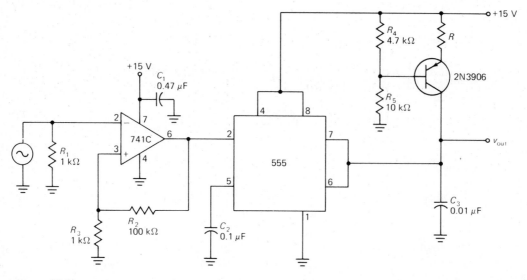

Figure 53-4

monostable operation. But now the timing capacitor is charged by a *pnp* current source rather than a resistor. For each value of R listed in Table 53-4, calculate the slope of the output waveform.

16. Connect the circuit of Fig. 53-4 with an R of 10 kΩ.

17. Set the ac generator to 1 kHz. Adjust the level to get a duty cycle of approximately 90 percent out of the Schmitt trigger.

18. Look at the output voltage; it should be a positive ramp. Measure the ramp voltage and time. Then work out the slope. Record the value in Table 53-4.

19. Repeat Steps 16 to 18 for the remaining values of R (Table 53-4).

Troubleshooting (Optional)

20. Assume that R equals 22 kΩ in Fig. 53-4. Here are the symptoms: (1) no ramp appears at the final output; (2) a normal Schmitt-trigger output drives pin 2 of the 555 timer; (3) approximately + 10 V appears at the base of the 2N3904. Try to figure out what troubles (there is more than one possibility) can cause these symptoms. Insert each suspected trouble to verify that it does cause the symptoms. Record all the troubles you locate (Table 53-5).

Design (Optional)

21. Select a value of C_3 that produces a slope of 15 V/ms when R equals 10 kΩ.

22. Connect the circuit with an R of 10 kΩ and your design value of C_3. Measure the slope of the output signal. Record the quantities of Table 53-6.

Computer (Optional)

23. Write and run a program that calculates the frequency and duty cycle of a 555 timer connected as an astable multivibrator.

DATA FOR EXPERIMENT 53

Table 53-1. Astable Multivibrator

		Calculated		Measured	
R_A	R_B	f	D	f	D
10 kΩ	100 kΩ				
100 kΩ	10 kΩ				
10 kΩ	10 kΩ				

Table 53-2. VCO Operation

$f_{min} = $ _____

$f_{max} = $ _____

Table 53-3. Monostable Multivibrator

R	Calculated W	Measured W
33 kΩ		
47 kΩ		
68 kΩ		

Table 53-4. Ramp Generator

R	Calculated Slope	Measured Slope
10 kΩ		
22 kΩ		
33 kΩ		

Table 53-5. Troubleshooting

Trouble	Description
1	
2	
3	
4	

Table 53-6. Design

$C_3 = $ _____

Slope $ = $ _____

QUESTIONS FOR EXPERIMENT 53

1. In Fig. 53-1, the calculated frequency for R_A and R_B both equal to 10 kΩ is ( )
   approximately:
   (a) 686 Hz;   (b) 1.2 kHz;   (c) 4.8 kHz;   (d) 6.91 kHz.

2. In Fig. 53-2, the adjustment controls the: ( )
   (a) output frequency;   (b) output voltage;   (c) supply voltage;   (d) input voltage

3. The output of the Schmitt trigger in Fig. 53-3 is: ( )
   (a) always positive;   (b) always negative;   (c) positive on one half-cycle
   and negative on the other;   (d) a constant dc voltage.

4. An R of 47 kΩ in Fig. 53-3 produces a pulse that is closest to: ( )
   (a) 363 μs;   (b) 517 μs;   (c) 748 μs;   (d) 1000 μs.

5. In Fig. 53-4, an R of 10 kΩ produces a slope of approximately: ( )
   (a) 12.4 V/ms;   (b) 18.6 V/ms;   (c) 41 V/ms;   (d) 56 V/ms.

6. In Fig. 53-3, how much input voltage do we need to get an output from the Schmitt
   trigger? Why?

7. Briefly describe the circuit operation of Fig. 53-4.

Troubleshooting (Optional)

8. In Fig. 53-4, assume R equals 10 kΩ and the output slope is 410 V/ms. Name a
   trouble that can produce this slope.

Design (Optional)

9. How did you figure out your design values?

10. Optional. Instructor's question.

EXPERIMENT 54

VOLTAGE REGULATION

▼

The dc voltage out of a bridge rectifier has a peak-to-peak ripple typically around 10 percent of the unregulated dc voltage. By using this unregulated voltage as the input to a voltage regulator, we can produce a dc output voltage that is almost constant with very small ripple. A voltage regulator uses noninverting voltage feedback. The input or reference voltage comes from a zener diode. This zener voltage is amplified by the closed-loop voltage gain of the regulator. The result is a larger dc output voltage with the same temperature coefficient as the zener diode. Most voltage regulators include current limiting to prevent an accidental short across the load terminals from destroying the pass transistor or diodes in the unregulated supply.

REQUIRED READING

Chapter 23 (Secs. 23-1 to 23-3) to *Electronic Principles*, 5th ed.

EQUIPMENT

- 1 power supply: adjustable from 0 to 15 V
- 11 ½-W resistors: 100 Ω, 220 Ω, 330 Ω, 470 Ω, two 680 Ω, 1 kΩ, two 2.2 kΩ, 4.7 kΩ, 10 kΩ
- 1 zener diode: 1N753
- 3 transistors: 2N3904
- 1 capacitor: 0.1 μF
- 1 VOM (analog or digital multimeter)

PROCEDURE

Minimum and Maximum Load Voltage

1. In Fig. 54-1, what is the approximate voltage across the zener diode? Record in Table 54-1. (Note: A bypass capacitor of 0.1 μF is used to prevent parasitic oscillations, an undesirable effect that is discussed in Chap. 22 of your textbook.)
2. When R_5 is varied, the load voltage changes in Fig. 54-1. Calculate and record the minimum and maximum load voltage.

3. Connect the circuit.
4. Adjust the dc input voltage V_S to $+15$ V. Measure and record the zener voltage.
5. Adjust R_5 to get the minimum load voltage. Measure and record $V_{L(\min)}$.
6. Adjust R_5 to get maximum load voltage. Measure and record $V_{L(\max)}$.

Load Regulation

7. Adjust R_5 to get a load voltage of 10 V. Record this as the no-load voltage in Table 54-2.
8. Connect a load resistance of 1 kΩ. Measure the load voltage. Record this as the full-load voltage in Table 54-2.
9. Calculate and record the percent load regulation.

Line Regulation

10. Measure the load voltage. Record this as $V_{L(\max)}$ under source regulation in Table 54-2.
11. Decrease the input voltage from $+15$ to $+12$ V. This represents a line change of approximately 20 percent. Measure and record the load voltage as $V_{L(\min)}$.
12. Return the input voltage to $+15$ V. Calculate and record the percent source regulation.

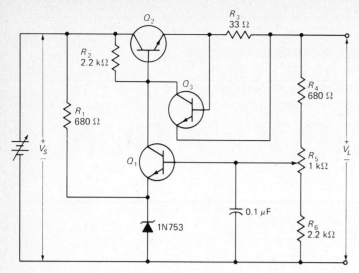

Figure 54-1

Current Limiting

13. Assume that the load voltage is 10 V and that Q_3 turns on when V_{BE} is 0.7 V. Notice that the R_4-R_5-R_6 voltage divider has some current through it. Calculate the load current where current limiting begins in Fig. 54-1. Record this at the top of Table 54-3.

14. Connect an R_L of 10 kΩ. Adjust the load voltage to 10 V. Then measure and record the load voltage for each load resistance listed in Table 54-3.

15. Connect a load resistance of 1 kΩ. Short the load terminals and notice how the load voltage goes to zero. Remove the short from the load terminals and notice how the load voltage returns to normal.

16. Use the VOM as an ammeter. Connect the VOM directly across the load terminals. It now measures the load current with a shorted load. This reading should be in the vicinity of your calculated I_{SL} at the top of Table 54-3.

Troubleshooting (Optional)

17. For each trouble listed in Table 54-4, estimate and record the load voltage.

18. Insert each trouble into the circuit. Measure and record the load voltage.

Design (Optional)

19. Select a value for R_4 to get a theoretical load voltage of approximately 9 to 13.2 V.
20. Connect the circuit with your value of R_4.
21. Measure the minimum and maximum load voltage. Record all quantities listed in Table 54-5.

Computer (Optional)

22. Write and run a program that calculates the minimum load voltage, maximum load voltage, and maximum load current in Fig. 54-1.

DATA FOR EXPERIMENT 54

Table 54-1. Minimum and Maximum Load Voltage

Calculated	Measured
$V_Z =$ _____	$V_Z =$ _____
$V_{L(min)} =$ _____	$V_{L(min)} =$ _____
$V_{L(max)} =$ _____	$V_{L(max)} =$ _____

Table 54-2. Regulation

Load Regulation	Source Regulation
$V_{NL} =$ _____	$V_{L(max)} =$ _____
$V_{FL} =$ _____	$V_{L(min)} =$ _____
%LR = _____	%SR = _____

Table 54-3. Current Limiting: $I_{SL} =$ _____

R_L	V_L
10 kΩ	
4.7 kΩ	
1 kΩ	
470 Ω	
330 Ω	
220 Ω	
100 Ω	
0	

Table 54-4. Troubleshooting

Trouble	Estimated V_{out}	Measured V_{out}
R_2 open		
Zener open		
Zener short		
Q_1 open		

Table 54-5. Design

$R_4 =$ _____

$V_{L(min)} =$ _____

$V_{L(max)} =$ _____

QUESTIONS FOR EXPERIMENT 54

1. The zener voltage of Fig. 54-1 is approximately: ( )
   (a) 5 V;   (b) 6.2 V;   (c) 7.5 V;   (d) 15 V.
2. Theoretically, the maximum regulated load voltage of Fig. 54-1 is approxi- ( )
   mately:
   (a) 6.2 V;   (b) 8.37 V;   (c) 12.2 V;   (d) 15 V.
3. Current limiting of Fig. 54-1 starts near: ( )
   (a) 1 mA;   (b) 2.25 mA;   (c) 12.5 mA;   (d) 18.6 mA.
4. The data of Table 54-2 show that load voltage is: ( )
   (a) dependent on load current;   (b) proportional to source voltage;
   (c) almost constant;   (d) low.
5. When V_L is +10 V and R_L is 1 kΩ in Fig. 54-1, the power dissipation in ( )
   the pass transistor is approximately:
   (a) 50 mW;   (b) 100 mW;   (c) 200 mW;   (d) 279 mW.
6. Briefly explain how the voltage regulator of Fig. 54-1 works.

7. Assume the load terminals are shorted in Fig. 54-1. If the source voltage is +15 V, what is the power dissipation in the pass transistor?

Troubleshooting (Optional)

8. Why does the load voltage approach the source voltage when the zener diode opens in Fig. 54-1?

Design (Optional)

9. Suppose we want current limiting to start at approximately 100 mA. What changes are necessary?

10. Optional. Instructor's question.

EXPERIMENT 55

THREE-TERMINAL IC REGULATORS

▼

The LM340 series is typical of the IC voltage regulators currently available. These three-terminal regulators are the ultimate in simplicity. And they are virtually indestructible because of the thermal shutdown discussed in your textbook. In this experiment you will connect an LM340-5 as a voltage regulator and a current regulator.

REQUIRED READING

Chapter 23 (Sec. 23-4) of *Electronic Principles*, 5th ed.

EQUIPMENT

1 audio generator
1 power supply: adjustable from 0 to 15 V
1 voltage regulator: LM340-5
6 ½-W resistors: 10 Ω, 22 Ω, 33 Ω, 47 Ω, 68 Ω, 150 Ω
2 capacitors: 0.1 μF, 0.22 μF
1 VOM (analog or digital multimeter)
1 oscilloscope

PROCEDURE

Voltage Regulator

1. In Fig. 55-1, estimate and record the output voltage for each input voltage listed in Table 55-1.
2. Connect the circuit.
3. Measure and record the output voltage for each input voltage listed in Table 55-1.

Ripple Rejection

4. In Fig. 55-2, the ac source in series with the dc source simulates ripple superimposed on dc voltage. The data sheet of an LM340-5 (see Appendix) lists the following ripple rejection: minimum is 62 dB and typical is 80 dB. Calculate the peak-to-peak ac output for the minimum and typical ripple rejection. Record your answers in Table 55-2.
5. Connect the circuit of Fig. 55-2.
6. Look at the ac voltage at the input to the regulator. Adjust the signal source to get 2 V p-p at 120 Hz.
7. Use the most sensitive ac ranges of the oscilloscope to look at the output ripple. Measure and record this ac output voltage. Then calculate and record the ripple rejection ratio in decibels.

Adjustable Voltage Regulator and Current Regulator

8. The circuit of Fig. 55-3 can function either as a voltage regulator if you use the output voltage or as a current regulator if R_2 is the load resistor. Calcu-

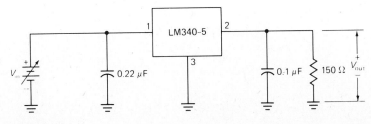

Figure 55-1

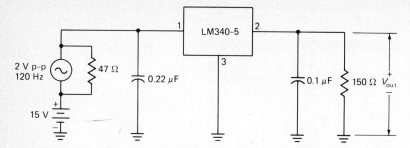

Figure 55-2

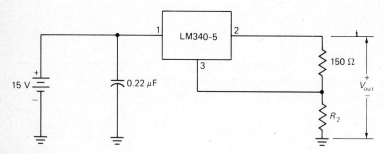

Figure 55-3

late and record V_{out} and I_{out} for each value of R_2 listed in Table 55-3.

9. Connect the circuit with an R_2 of 10 Ω. Measure and record V_{out} and I_{out}.

10. Repeat Step 9 for the other values of R_2.

Troubleshooting (Optional)

11. Assume the circuit of Fig. 55-3 has an R_2 of 68 Ω. For each trouble in Table 55-4, estimate and record the dc output voltage.

12. Connect the circuit with an R_2 of 68 Ω. Insert each trouble. Measure and record the dc output voltage.

Design (Optional)

13. Select a value of R_2 in Fig. 55-3 to produce an output voltage of approximately 9 V.

14. Insert your value of R_2. Measure the output voltage. Record R_2 and V_{out} in Table 55-5.

Computer (Optional)

15. Write and run a program that calculates the dc output voltage in Fig. 55-3.

DATA FOR EXPERIMENT 55

Table 55-1. Voltage Regulator

V_{in}	Estimated V_{out}	Measured V_{out}
1 V		
5 V		
10 V		
11 V		
12 V		
13 V		
14 V		
15 V		

Table 55-2. Ripple Rejection

Calculated V_{rip} (minimum rejection) =

Calculated V_{rip} (typical rejection) =

Measured V_{rip} =

Measured ripple rejection =

Table 55-3. Voltage and Current Regulation

R_2	Calculated V_{out}	I_{out}	Measured V_{out}	I_{out}
10 Ω				
22 Ω				
33 Ω				
47 Ω				
68 Ω				

Table 55-4. Troubleshooting

Trouble	Estimated V_{out}	Measured V_{out}
R_1 short		
R_1 open		
R_2 short		
R_2 open		

Table 55-5. Design

$R_2 =$ _____

$V_{out} =$ _____

QUESTIONS FOR EXPERIMENT 55

1. When the input voltage of Fig. 55-1 is greater than 10 V, the output voltage ( )
   is approximately:
   (a) constant;   (b) 5 V;   (c) regulated;   (d) all of the foregoing.
2. In Table 55-2, the typical output ripple is approximately: ( )
   (a) 0.2 mV;   (b) 1 mV;   (c) 1.59 mV;   (d) 10 mV.
3. If I_Q is 8 mA in Fig. 55-3, the calculated I_{out} is approximately: ( )
   (a) 8 mA;   (b) 23.3 mA;   (c) 41.3 mA;   (d) 100 mA.
4. When R_2 is 68 Ω in Fig. 55-3, the calculated V_{out} is approximately: ( )
   (a) 5.43 V;   (b) 6.43 V;   (c) 7.04 V;   (d) 7.94 V.
5. The measured current in Table 55-3 indicates that the regulator circuit can ( )
   function as a:
   (a) current source;   (b) voltage source;   (c) ripple generator;   (d) amplifier.
6. Briefly explain why the data sheet of an LM340-5 indicates that the input voltage
   must be at least 7 V.

7. Why are bypass capacitors used with an IC regulator?

Troubleshooting (Optional)

8. What output voltage did you get when R_1 was open? Why did you get this voltage?

Design (Optional)

9. What value did you use for R_2 to get an output of 9 V? How did you arrive at this
   value?

10. Optional. Instructor's question.

EXPERIMENT 56

CLASS C AMPLIFIERS

▼

In a class C amplifier, the transistor operates in the active region for less than 180° of the ac cycle. Typically, the conduction angle is much smaller than 180° and the collector current is a train of narrow pulses. This highly nonsinusoidal current contains a fundamental frequency plus harmonics. A tuned class C amplifier has a resonant tank circuit that is tuned to the fundamental frequency. This produces a sinusoidal output voltage of frequency f_r. In a frequency multiplier, the resonant tank circuit is tuned to the nth harmonic, so that the sinusoidal output has a frequency of nf_r.

In this experiment you will build a tuned class C amplifier and a frequency multiplier.

REQUIRED READING

Chapter 24 (Secs. 24-1 to 24-3) of *Electronic Principles*, 5th ed.

EQUIPMENT

1 audio generator
1 power supply: 10 V
1 transistor: 2N3904
3 ½-W resistors: 220 Ω, two 100 kΩ
1 inductor: 15 mH (J. W. Miller 70F152AI or equivalent)
3 capacitors: 0.0047 µF, two 1 µF
1 VOM (analog or digital multimeter)
1 oscilloscope
1 frequency counter

PROCEDURE

Untuned Class C Amplifier

1. Connect the circuit of Fig. 56-1.
2. Set the input frequency of 20 kHz. Adjust the signal level to get narrow output pulses with a peak-to-peak value of 6 V.
3. Measure the pulse width W and period T. Record these values in Table 56-1. Calculate and record the duty cycle.
4. Look at the signal on the base. You should see a negatively clamped waveform. With the oscilloscope

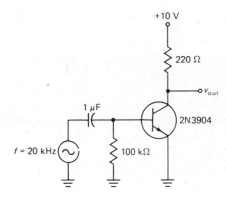

Figure 56-1

on dc input, measure and record the positive and negative peak voltages of this clamped signal.
5. In Fig. 56-2, calculate the resonant frequency of the tuned amplifier. Record your answer in Table 56-2.
6. Assume the Q_L of the coil is 15. Calculate and record the other quantities listed in Table 56-2.
7. Connect the circuit. Use the oscilloscope to look at the base signal. Adjust the frequency to 20 kHz and the signal level to 2 V p-p.
8. Use the oscilloscope to look at the collector signal. Vary the input frequency until the output signal reaches a maximum value (resonance).

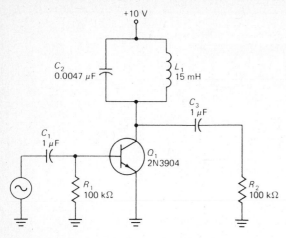

Figure 56-2

9. Adjust the signal level as needed to get 15 V p-p at the collector.

10. Repeat Steps 8 and 9 until the circuit is resonant with an output of 15 V p-p. (This repetitive adjustment of frequency and signal level is called "rocking in.")

Resonant Frequency, Bandwidth, and Circuit Q

11. Use the frequency counter to measure the resonant frequency. Record in Table 56-3.

12. Measure and record the bandwidth. If you don't know how to do this, look at Eq. (24-4) in your textbook and figure out how to do it. After measuring f_1 and f_2, calculate and record the experimental bandwidth in Table 56-3.

13. Readjust the input frequency to get resonance. Calculate the ciruit Q using the f_r and B of Table 56-3. Record the circuit Q.

AC Output Compliance, Current Drain, and DC Clamping

14. Increase the input signal level until the output signal just starts clipping. Back off slightly from this level until the signal stops clipping. Record the ac output compliance in Table 56-3.

15. Use the VOM as an ammeter to measure the current drain of the circuit. Record I_S.

Frequency Multiplier

16. Reduce the input frequency and notice that the output signal decreases (off resonance). Continue decreasing the input frequency until the output signal again reaches a maximum value (resonant to a harmonic). Use the frequency counter to measure the input and output frequencies. Record in Table 56-4. Divide f_{out} by f_{in}, round the answer off to the nearest integer, and notice it equals 2. The circuit is now operating as an X2 frequency multiplier.

17. Again decrease the input frequency until you find another resonance. Measure and record the input and output frequencies. This time, the f_{out}/f_{in} ratio should be approximately 3. The circuit is now acting like an X3 frequency multiplier.

Troubleshooting (Optional)

18. In Fig. 56-2, assume R_1 is open. Estimate the ac load voltage for this trouble. Record in Table 56-5.

19. Repeat Step 18 for each trouble listed in Table 56-5.

20. Connect the circuit with each trouble. Measure and record the ac load voltage.

Design (Optional)

21. Select a value of C_2 (nearest standard value) to get a resonant frequency of approximately 12 kHz.

22. Connect the circuit with your design value of C_2. Tune to the fundamental frequency. Record the capacitance and frequency here:

$$C_2 = \underline{\hspace{2cm}}$$

$$f_r = \underline{\hspace{2cm}}$$

Computer (Optional)

23. Enter and run this program:

```
  10 PRINT "ENTER A NUMBER FROM 1 to
     3": INPUT X
  20 ON X GOSUB 1000, 2000, 3000
  30 GOTO 10
1000 PRINT "YOU SELECTED 1": RETURN
2000 PRINT "YOU SELECTED 2": RETURN
3000 PRINT "YOU SELECTED 3": RETURN
```

24. Write and run a program that prints out the resonant frequency, bandwidth, and ac output compliance of Fig. 56-2. The inputs are L, C, Q, and V_{CC}.

DATA FOR EXPERIMENT 56

Table 56-1. Waveforms

$W =$ _____

$T =$ _____

$D =$ _____

$+ \text{ peak} =$ _____

$- \text{ peak} =$ _____

Table 56-2. Calculations for Tuned Amplifier

$f_r =$ _____

$X_L =$ _____

$R_S =$ _____

$R_P =$ _____

$r_C =$ _____

$Q =$ _____

$B =$ _____

$PP =$ _____

$P_{L(\text{max})} =$ _____

Table 56-3. Measurements for Tuned Amplifier

$f_r =$ _____

$B =$ _____

$Q =$ _____

$PP =$ _____

$I_S =$ _____

Table 56-4. Frequency Multiplier

f_{in}	f_{out}	n
		2
		3

Table 56-5. Troubleshooting

Trouble	Estimated v_{out}	Measured v_{out}
R_1 open		
Q_1 collector-emitter short		
C_2 short		
C_3 open		

QUESTIONS FOR EXPERIMENT 56

1. The duty cycle of Table 56-1 is closest to: ( )
   (a) 1%;   (b) 10%;   (c) 31.6%;   (d) 75%.
2. The ac output compliance of Table 56-2 is approximately: ( )
   (a) 0.7 V;   (b) 1.4 V;   (c) 10 V;   (d) 20 V.
3. To calculate the total dc input power to the tuned amplifier, we can multiply   ( )
   the measured current drain of Table 56-3 by the:
   (a) ac output compliance;   (b) circuit Q;   (c) supply voltage;
   (d) bandwidth.
4. Assume the current drain equals 0.25 mA in Fig. 56-2. If the ac output   ( )
   compliance is 19.6 V, then the stage efficiency is approximately:
   (a) 5%;   (b) 19%;   (c) 47%;   (d) 73%.
5. When f_{in} = 4.75 kHz and f_{out} = 19 kHz, a frequency multiplier is tuned to   ( )
   which harmonic of the fundmental frequency?
   (a) First;   (b) second;   (c) third;   (d) fourth.

6. Briefly explain how a tuned class C amplifier works.

7. Explain how a frequency multiplier works.

Troubleshooting (Optional)

8. The input voltage driving the tuned class C amplifier of Fig. 56-2 is 1 V p-p. The output voltage is zero. Name the most likely trouble.

9. In this experiment, the tuned class C amplifier has a stage efficiency of only 20 percent, more or less depending on the components used. What do you think was the cause of this poor efficiency? Explain your answer.

10. Optional. Instructor's question.

EXPERIMENT 57

THE FREQUENCY MIXER

▼

In a frequency mixer, two input signals drive a nonlinear device to produce the original frequencies, the sum and difference frequencies, and other intermodulation frequencies. Typically, the output of the mixer is filtered to get only the difference frequency. For a mixer to work properly, one of the input signals must be large enough to produce large-signal operation; this ensures the nonlinear operation needed to produce intermodulation frequencies. The other input signal is usually small, so that the amplitude of the difference-frequency output is proportional to the amplitude of the small input signal.

REQUIRED READING

Chapter 24 (Secs. 24-4 to 24-5) of *Electronic Principles*, 5th ed.

EQUIPMENT

2 signal generators: variable in the vicinity of 50 kHz
1 power supply: 15 V
1 transistor: 2N3904
9 ½-W resistors: 100 Ω, two 10 kΩ, two 15 kΩ, two 22 kΩ, two 33 kΩ, two 68 kΩ
7 capacitors: two 100 pF, two 220 pF, 1000 pF, 0.1 μF, 1 μF
1 VOM (analog or digital multimeter)

1 oscilloscope
1 frequency counter

PROCEDURE

Calculations

1. Figure 57-1 shows a bipolar mixer where v_x is the small signal and v_y is the large signal. If f_x is 51 kHz and f_y is 50 kHz, calculate and record the group-1 frequencies in Table 57-1.
2. Calculate and record the group-2 frequencies.
3. Calculate and record the group-3 frequencies.
4. All the foregoing frequencies plus other intermodulation components will appear at the collector of Fig.

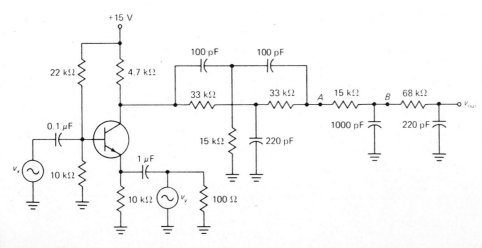

Figure 57-1

57-1. The twin-T filter has a notch of approximately 50 kHz, so it will greatly attenuate signals with frequencies of f_x and f_y. In addition, each lag network has a cutoff frequency of approximately 10 kHz. Under ideal conditions, if the filters stop all frequencies above 10 kHz, which of the frequencies recorded in Steps 1 to 3 appear at the final output? Record your answers in Table 57-2.

Mixer Circuit

5. Connect the circuit of Fig. 57-1.
6. Reduce v_x to zero. Use the oscilloscope to look at the collector voltage of the 2N3904. Set f_y to approximately 50 kHz. Adjust the v_y generator to produce 6 V p-p at the collector. (The signal will appear clipped.)
7. Use the oscilloscope to look at the v_x input signal. Set the f_x to approximately 51 kHz. Adjust the v_x generator to get an input of 20 mV p-p.
8. Look at the final output signal with a vertical sensitivity of 0.1 V/cm (ac input) and a sweep time of 1 ms/cm. Vary the frequency of the v_x generator until you get an output of 1 kHz.
9. Look at point B, the input to the final lag network. Change the sweep speed to 0.1 ms/cm. Notice the ripple on the 1-kHz signal.
10. Look at point A, the input to the first lag network. Notice how large the ripple is here.
11. Return to the final output. If the frequency has

drifted, readjust f_x to get an output of 1 kHz. Measure and record the peak-to-peak voltage in Table 57-3.
12. Calculate the conversion voltage gain. Record in Table 57-3.

Troubleshooting (Optional)

13. Ask the instructor to insert a trouble in your circuit.
14. Locate and repair the trouble. Record the trouble in Table 57-4.
15. Repeat Steps 13 and 14 as often as indicated by the instructor.

Design (Optional)

16. By changing the operating point, you can get more conversion voltage gain. There are other changes you can also try like adjusting the signal level of the large signal. Optimize the circuit to get maximum conversion voltage gain.
17. Measure the conversion voltage gain for a small-signal input of 20 mV p-p. Record your changes and the new conversion voltage gain in Table 57-5.

Computer (Optional)

18. Write and run a program that calculates the frequency components in groups 1, 2, and 3 of a large-signal amplifier.

DATA FOR EXPERIMENT 57

Table 57-1. Calculations

Group 1:

$f_x - f_y$ = _____ kHz & _____ kHz

$f_x - 2f_y$ = _____ kHz & _____ kHz

$f_x - 3f_y$ = _____ kHz & _____ kHz

Group 2:

$2f_x - f_y$ = _____ kHz & _____ kHz

$2f_x - 2f_y$ = _____ kHz & _____ kHz

$2f_x - 3f_y$ = _____ kHz & _____ kHz

Group 3:

$3f_x - f_y$ = _____ kHz & _____ kHz

$3f_x - 2f_y$ = _____ kHz & _____ kHz

$3f_x - 3f_y$ = _____ kHz & _____ kHz

Table 57-2. Output Components

f_1 = _____

f_2 = _____

f_3 = _____

Table 57-3. Conversion Gain

v_{out} = _____

Conversion voltage gain = _____

Table 57-4. Troubleshooting

Trouble	Description
1	
2	
3	

Table 57-5. Design

Change	Description
1	
2	

Conversion voltage gain = _____

QUESTIONS FOR EXPERIMENT 57

1. The lowest frequency in group 1 of Table 57-1 is: (   )
   (a) 1 kHz;   (b) 2 kHz;   (c) 49 kHz;   (d) 50 kHz.
2. The highest frequency in group 1 of Table 57-1 is: (   )
   (a) 99 kHz;   (b) 101 kHz;   (c) 151 kHz;   (d) 201 kHz.
3. The lowest frequency in group 3 is: (   )
   (a) 1 kHz;   (b) 3 kHz;   (c) 53 kHz;   (d) 103 kHz.
4. The measured conversion voltage gain of this experiment was closest to: (   )
   (a) −20 dB;   (b) −10 dB;   (c) 20 dB;   (d) 40 dB.
5. The ripple on the signal at point B is approximately: (   )
   (a) 2 kHz;   (b) 3 kHz;   (c) 10 kHz;   (d) 50 kHz.
6. Explain how the circuit of Fig. 57-1 works.

Troubleshooting (Optional)

7. Suppose the final output of Fig. 57-1 contains a large signal with a frequency of 50 kHz. What part of the circuit would you suspect?

8. The dc voltage at the final output of Fig. 57-1 is normal, but the difference signal is very weak. Name at least three troubles that could cause this.

Design (Optional)

9. What kind of a mixer would you use to reduce the harmonics and spurious signals? Why does this help?

10. Optional. Instructor's question.

EXPERIMENT 58

AMPLITUDE MODULATION AND DEMODULATION

▼

With amplitude modulation, a low-frequency signal controls the amplitude of a high-frequency carrier. If the modulating signal is sinusoidal, then the envelope of the modulated RF signal will also be sinusoidal. Amplitude modulation creates an upper and lower sideband. In some communication systems, the carrier or one of the sidebands may be suppressed.

After an AM signal has been received, it can be demodulated to recover the low-frequency signal. One of the simplest enveloped detectors is a diode peak detector. The time constant of the peak detector is long compared with the RF period, but short compared with the modulating period.

REQUIRED READING

Chapter 24 (Secs. 24-7 to 24-10) of *Electronic Principles*, 5th ed.

EQUIPMENT

1 audio generator
1 RF generator
1 power supply: 15 V
1 diode: 1N914
1 transistor: 2N3904
10 ½-W resistors: 470 Ω, four 1 kΩ, 2.2 kΩ, three 10 kΩ, 22 kΩ
4 capacitors: two 0.001 μF, 0.01 μF, 0.1 μF
1 oscilloscope

PROCEDURE

Modulated RF Amplifier

1. Assume the audio voltage is zero in Fig. 58-1. What is the ideal voltage gain for the carrier in Fig. 58-1? Record the approximate value in Table 58-1. (Note: The 470-Ω resistor is a dc return in case your audio generator is ac-coupled. You can omit this component if your generator is dc-coupled.)

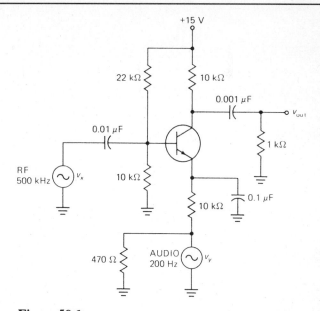

Figure 58-1

2. If the audio signal has a peak-to-peak voltage of 6 V, what are the minimum and maximum voltage gains for the carrier? Record your answers.
3. With the foregoing gains, calculate and record the percent modulation.

4. Connect the circuit of Fig. 58-1.
5. Set the audio generator to 200 Hz and the RF generator to 500 kHz.
6. Reduce the audio signal level to zero (do not disconnect). Adjust the RF generator to get a final output v_{out} of 0.3 V p-p (unmodulated).
7. Measure the RF input signal v_x and record the peak-to-peak value in Table 58-2. Calculate and record the value of A_0.
8. Look at v_{out} with a sensitivity of 0.1 V/cm and a sweep speed at 1 ms/cm. Increase the audio signal and you will see amplitude modulation.
9. Look at v_y and set the level at 6 V p-p.
10. Look at v_{out} and record the maximum and minimum peak-to-peak values (these are $2V_{max}$ and $2V_{min}$). Then calculate and record the percent modulation.
11. Increase and decrease the audio level and notice how the percent modulation changes.

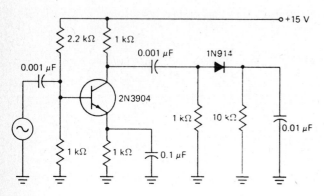

Figure 58-2

Envelope Detector

12. Connect the circuit of Fig. 58-2.
13. Set the RF generator to 500 kHz unmodulated. Look at the input to the envelope detector. Adjust the RF generator to get a signal level of 5 V p-p with a sweep time of 1 ms/cm.
14. Turn on the modulation. If it is adjustable, set the modulation at 30 percent.
15. With the oscilloscope at 0.1 V/cm, look at the output of the envelope detector. You should have an audio signal.

Troubleshooting (Optional)

16. Ask the instructor to insert a trouble in the circuit of Fig. 58-1.
17. Locate and repair the trouble. Record the trouble in Table 58-3.
18. Repeat Steps 16 and 17 as often as indicated by the instructor.

Design (Optional)

19. Select a value of load resistance to get an unmodulated voltage gain of approximately 30.
20. Connect the circuit with your design value of load resistance. Measure the unmodulated voltage gain.
21. Record the quantities of Table 58-4.

Computer (Optional)

22. Write and run a program that calculates percent modulation. The inputs are A_{max} and A_{min}.

266

DATA FOR EXPERIMENT 58

Table 58-1. Calculations

$A_0 =$ _____

$A_{max} =$ _____

$A_{min} =$ _____

Percent $m =$ _____

Table 58-2. Measurements

$v_x =$ _____

$A_0 =$ _____

$2V_{max} =$ _____

$2V_{min} =$ _____

Percent $m =$ _____

Table 58-3. Troubleshooting

Trouble	Description
1	
2	
3	

Table 58-4. Design

$R_L =$ _____

$A_0 =$ _____

QUESTIONS FOR EXPERIMENT 58

1. The unmodulated RF voltage gain of Table 58-1 is closest to: ( )
   (a) 1; (b) 5.6; (c) 9.8; (d) 14.5.
2. The calculated percent modulation of Table 58-1 is closest to: ( )
   (a) 10%; (b) 53%; (c) 75%; (d) 90%.
3. In Fig. 58-1, the emitter bypass capacitor appears like an ac short to which ( )
   of these signals?
   (a) audio; (b) RF; (c) supply voltage; (d) emitter voltage.
4. The envelope detector of Fig. 58-2 is: ( )
   (a) an average detector; (b) a peak detector; (c) an RF amplifier;
   (d) an amplitude modulator.
5. If the percent modulation is 30% in Fig. 58-2, the highest unattenuated ( )
   frequency the enveloped detector can produce is:
   (a) 1 kHz; (b) 5.31 kHz; (c) 7.83 kHz; (d) 10 kHz.

6. Explain how the circuit of Fig. 58-1 works.

7. Explain how the envelope detector of Fig. 58-2 works.

Troubleshooting (Optional)

8. All dc voltages are normal in Fig. 58-1. The unmodulated RF output signal is very weak. Name three possible troubles.

Design (Optional)

9. In Fig. 58-1, the audio generator is capacitively coupled. Why is it necessary to use a dc return?

10. Optional. Instructor's question.

APPENDIX

PARTS AND EQUIPMENT

Resistors (all ½ W unless otherwise specified)

Quantity	Description
1	10 Ω
2	22 Ω
1	47 Ω
1	51 Ω
2	68 Ω
2	100 Ω
1	100 Ω, 1 W
2	150 Ω
1	180 Ω
2	220 Ω
2	270 Ω, 1 W
2	330 Ω
4	470 Ω
1	470 Ω, 1 W
1	560 Ω
2	680 Ω
1	750 Ω
1	910 Ω
4	1 kΩ
1	1.1 kΩ
1	1.2 kΩ
2	1.5 kΩ
1	1.8 kΩ
1	2 kΩ
2	2.2 kΩ
3	3.6 kΩ
1	4.3 kΩ
2	4.7 kΩ
2	6.8 kΩ
1	7.5 kΩ
1	8.2 kΩ
3	10 kΩ
2	15 kΩ
1	18 kΩ
2	22 kΩ
1	24 kΩ
2	33 kΩ
1	36 kΩ
2	47 kΩ
1	68 kΩ
2	100 kΩ
1	200 kΩ
2	220 kΩ
1	270 kΩ
1	330 kΩ
1	430 Ω
1	470 kΩ

Potentiometers

Quantity	Description
1	1 kΩ
1	5 kΩ

Capacitors (25-V rating or better)

Quantity	Description
2	100 pF
2	220 pF
2	1000 pF
1	2000 pF
1	0.0047 μF
2	0.01 μF
1	0.022 μF
2	0.047 μF
2	0.1 μF
1	0.22 μF
2	0.47 μF
3	1 μF
2	10 μF
2	47 μF
1	100 μF
2	470 μF

Diodes

Quantity	Description
3	1N753 (6.2-V zener)
1	1N757 (9-V zener)
3	1N914 (small-signal diode)
4	1N4001 (rectifier)
1	TIL221 (red LED)
1	TIL222 (green LED)

Transistors

Quantity	Description
3	2N3904 (small-signal *npn*)
3	2N3906 (small-signal *pnp*)
1	2N3055 (power *npn*)
1	2N4444 (SCR)
3	MPF102 (*n*-channel JFET)
3	VN10KM (Radio Shack VMOS transistor)

Integrated Circuits

Quantity	Description
1	LM318C (op amp)
3	LM741C (op amp)
1	LM340-5 (voltage regulator)
1	NE555 (timer)
1	NE565 (phase-locked loop)

Equipment

Quantity	Description
1	AC voltmeter (with decibel scale)
1	Audio generator
1	Frequency counter
1	Milliammeter (or second VOM)
1	Oscilloscope

Quantity	Description
2	Power supply: adjustable from 0 to 15 V
2	RF signal generator
1	Sine/square generator
1	VOM (analog or digital)

Miscellaneous

Quantity	Description
1	4N26 (optocoupler)
1	Inductor: 100 μH
1	Inductor: 15 mH
1	Switch: SPST
1	TIL312 (seven-segment display)
1	Transformer: 12.6 V ac center-tapped (Triad F-25X or equivalent) with fused line cord

DATA SHEETS

MOTOROLA Semiconductors
BOX 20912 • PHOENIX, ARIZONA 85036

Designers▲Data Sheet

500-MILLIWATT HERMETICALLY SEALED GLASS SILICON ZENER DIODES

- Complete Voltage Range — 2.4 to 91 Volts
- DO-35 Package — Smaller than Conventional DO-7 Package
- Double Slug Type Construction
- Metallurgically Bonded Construction
- Nitride Passivated Die

Designer's Data for "Worst Case" Conditions

The Designers▲ Data sheets permit the design of most circuits entirely from the information presented. Limit curves — representing boundaries on device characteristics — are given to facilitate "worst case" design.

MAXIMUM RATINGS

Rating	Symbol	Value	Unit
DC Power Dissipation @ $T_L \leq 50^oC$, Lead Length = 3/8''	P_D		
*JEDEC Registration		400	mW
*Derate above $T_L = 50^oC$		3.2	mW/^oC
Motorola Device Ratings		500	mW
Derate above $T_L = 50^oC$		3.33	mW/^oC
Operating and Storage Junction Temperature Range	T_J, T_{stg}		^oC
*JEDEC Registration		–65 to +175	
Motorola Device Ratings		–65 to +200	

*Indicates JEDEC Registered Data.

MECHANICAL CHARACTERISTICS

CASE: Double slug type, hermetically sealed glass

MAXIMUM LEAD TEMPERATURE FOR SOLDERING PURPOSES: 230^oC, 1/16'' from case for 10 seconds

FINISH: All external surfaces are corrosion resistant with readily solderable leads.

POLARITY: Cathode indicated by color band. When operated in zener mode, cathode will be positive with respect to anode.

MOUNTING POSITION: Any

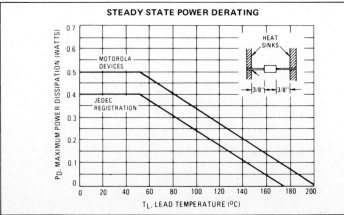

STEADY STATE POWER DERATING

▲Trademark of Motorola Inc.

1N746 thru 1N759

1N957 thru 1N984

1N4370 thru 1N4372

GLASS ZENER DIODES
500 MILLIWATTS
2.4-91 VOLTS

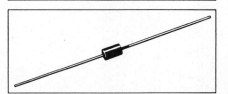

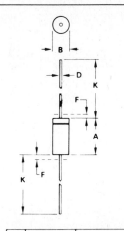

DIM	MILLIMETERS		INCHES	
	MIN	MAX	MIN	MAX
A	3.05	5.08	0.120	0.200
B	1.52	2.29	0.060	0.090
D	0.46	0.56	0.018	0.022
F	—	1.27	—	0.050
K	12.70	—	0.500	—

All JEDEC dimensions and notes apply.

CASE 299-01
DO-35

NOTE:
1. POLARITY DENOTED BY CATHODE BAND.
2. LEAD DIAMETER IS NOT CONTROLLED WITHIN DIMENSION "F"

©MOTOROLA INC. 1977   DS 7021 R2

ELECTRICAL CHARACTERISTICS ($T_A = 25^{\circ}C$, $V_F = 1.5$ V max at 200 mA for all types)

Type Number (Note 1)	Nominal Zener Voltage V_Z @ I_{ZT} (Note 2) Volts	Test Current I_{ZT} mA	Maximum Zener Impedance Z_{ZT} @ I_{ZT} (Note 3) Ohms	*Maximum DC Zener Current I_{ZM} (Note 4) mA		Maximum Reverse Leakage Current $T_A = 25^{\circ}C$ I_R @ $V_R = 1$ V μA	$T_A = 150^{\circ}C$ I_R @ $V_R = 1$ V μA
1N4370	2.4	20	30	150	190	100	200
1N4371	2.7	20	30	135	165	75	150
1N4372	3.0	20	29	120	150	50	100
1N746	3.3	20	28	110	135	10	30
1N747	3.6	20	24	100	125	10	30
1N748	3.9	20	23	95	115	10	30
1N749	4.3	20	22	85	105	2	30
1N750	4.7	20	19	75	95	2	30
1N751	5.1	20	17	70	85	1	20
1N752	5.6	20	11	65	80	1	20
1N753	6.2	20	7	60	70	0.1	20
1N754	6.8	20	5	55	65	0.1	20
1N755	7.5	20	6	50	60	0.1	20
1N756	8.2	20	8	45	55	0.1	20
1N757	9.1	20	10	40	50	0.1	20
1N758	10	20	17	35	45	0.1	20
1N759	12	20	30	30	35	0.1	20

Type Number (Note 1)	Nominal Zener Voltage V_Z (Note 2) Volts	Test Current I_{ZT} mA	Maximum Zener Impedance (Note 3) Z_{ZT} @ I_{ZT} Ohms	Z_{ZK} @ I_{ZK} Ohms	I_{ZK} mA	*Maximum DC Zener Current I_{ZM} (Note 4) mA		Maximum Reverse Current I_R Maximum μA	Test Voltage Vdc 5%	V_R 10%
1N957	6.8	18.5	4.5	700	1.0	47	61	150	5.2	4.9
1N958	7.5	16.5	5.5	700	0.5	42	55	75	5.7	5.4
1N959	8.2	15	6.5	700	0.5	38	50	50	6.2	5.9
1N960	9.1	14	7.5	700	0.5	35	45	25	6.9	6.6
1N961	10	12.5	8.5	700	0.25	32	41	10	7.6	7.2
1N962	11	11.5	9.5	700	0.25	28	37	5	8.4	8.0
1N963	12	10.5	11.5	700	0.25	26	34	5	9.1	8.6
1N964	13	9.5	13	700	0.25	24	32	5	9.9	9.4
1N965	15	8.5	16	700	0.25	21	27	5	11.4	10.8
1N966	16	7.8	17	700	0.25	19	37	5	12.2	11.5
1N967	18	7.0	21	750	0.25	17	23	5	13.7	13.0
1N968	20	6.2	25	750	0.25	15	20	5	15.2	14.4
1N969	22	5.6	29	750	0.25	14	18	5	16.7	15.8
1N970	24	5.2	33	750	0.25	13	17	5	18.2	17.3
1N971	27	4.6	41	750	0.25	11	15	5	20.6	19.4
1N972	30	4.2	49	1000	0.25	10	13	5	22.8	21.6
1N973	33	3.8	58	1000	0.25	9.2	12	5	25.1	23.8
1N974	36	3.4	70	1000	0.25	8.5	11	5	27.4	25.9
1N975	39	3.2	80	1000	0.25	7.8	10	5	29.7	28.1
1N976	43	3.0	93	1500	0.25	7.0	9.6	5	32.7	31.0
1N977	47	2.7	105	1500	0.25	6.4	8.8	5	35.8	33.8
1N978	51	2.5	125	1500	0.25	5.9	8.1	5	38.8	36.7
1N979	56	2.2	150	2000	0.25	5.4	7.4	5	42.6	40.3
1N980	62	2.0	185	2000	0.25	4.9	6.7	5	47.1	44.6
1N981	68	1.8	230	2000	0.25	4.5	6.1	5	51.7	49.0
1N982	75	1.7	270	2000	0.25	1.0	5.5	5	56.0	54.0
1N983	82	1.5	330	3000	0.25	3.7	5.0	5	62.2	59.0
1N984	91	1.4	400	3000	0.25	3.3	4.5	5	69.2	65.5

*Left column based upon JEDEC Registration, right column based upon Motorola rating.

NOTE 1. TOLERANCE AND VOLTAGE DESIGNATION

Tolerance Designation

The type numbers shown have tolerance designations as follows:

1N4370 series: ±10%, suffix A for ±5% units.
1N746 series: ±10%, suffix A for ±5% units.
1N957 series: ±20%, suffix A for ±10% units, suffix B for ±5% units.

NOTE 2. ZENER VOLTAGE (V_Z) MEASUREMENT

Nominal zener voltage is measured with the device junction in thermal equilibrium at the lead temperature of $30^{\circ}C \pm 1^{\circ}C$ and 3/8'' lead length.

NOTE 3. ZENER IMPEDANCE (Z_Z) DERIVATION

Z_{ZT} and Z_{ZK} are measured by dividing the ac voltage drop across the device by the ac current applied. The specified limits are for $I_Z(ac) = 0.1\ I_Z(dc)$ with the ac frequency = 60 Hz.

NOTE 4. MAXIMUM ZENER CURRENT RATINGS (I_{ZM})

Maximum zener current ratings are based on the maximum voltage of a 10% 1N746 type unit or a 20% 1N957 type unit. For closer tolerance units (10% or 5%) or units where the actual zener voltage (V_Z) is known at the operating point, the maximum zener current may be increased and is limited by the derating curve.

MAXIMUM RATINGS (25°C) [Note 1]

WIV	Working Inverse Voltage	20 V
I_O	Average Rectified Current	50 mA
I_F	Forward Current Steady State d.c.	75 mA
I_F	Recurrent Peak Forward Current	150 mA
$i_{f(surge)}$	Peak Forward Surge Current Pulse Width of 1 Second	500 mA
$i_{f(surge)}$	Peak Foward Surge Current Pulse Width of 1 μs	2000 mA
P	Power Dissipation	250 mW
P	Power Dissipation	100 mW at 125°C
T_A	Operating Temperature	−65° to +175°C
T_{stg}	Storage Temperature, Ambient	−65° to +175°C

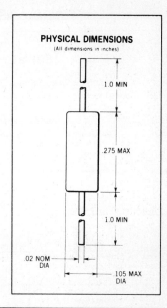

PHYSICAL DIMENSIONS
(All dimensions in inches)

1.0 MIN

.275 MAX

1.0 MIN

.02 NOM DIA

.105 MAX DIA

ELECTRICAL SPECIFICATIONS (25°C unless otherwise noted)

SYMBOL	CHARACTERISTIC	MIN.	TYP.	MAX.	TEST CONDITIONS
V_F	Forward Voltage			1.0 V	$I_F = 10$ mA
I_R	Reverse Current			25 NA	$V_R = -20$ V
I_R	Reverse Current (150°C)			50 μA	$V_R = -20$ V
BV	Breakdown Voltage	75 V		50 μA	$I_R = -25$ μA
BV	Breakdown Voltage	100V			$I_R = 100$ μA
t_{rr} (Note 2)	Reverse Recovery Time			4.0 ns	$I_F = 10$ mA $V_R = 6$ V
V_F (Note 5)	Peak forward recovery voltage			2.5 V	$I_F = 50$ mA pulse
C_O (Note 3)	Capacitance			4.0 pF	$V_R = 0V$ f = 1
R_E (Note 4)	Rectification Efficiency 45%				f = 100 MHz
$\Delta V_F/°C$	Forward Voltage Temperature Coefficient		−1.8mV/°C		

*Planar is a patented Fairchild process.

NOTES:
(1) The maximum ratings are limiting values above which life or satisfactory performance may be impaired.
(2) Recovery to 1 mA.
(3) Capacitance as measured on Boonton Electronic Corporation Model No. 75A-S8 Capacitance Bridge or equivalent.
(4) Rectification efficiency is defined as the ratio of D.C. load voltage to peak rf input voltage to the detector circuit, measured with 2.0 V r.m.s. input to the circuit. Load resistance 5k ohms, load capacitance 20 pF.
(5) Pulse width = 0.1 μs; Rise time of pulse equal to or less than 25 ns. Repetition rate 5 - 100 kHz.

FAIRCHILD
SEMICONDUCTOR
A DIVISION OF FAIRCHILD CAMERA AND INSTRUMENT CORPORATION

313 FAIRCHILD DRIVE, MOUNTAIN VIEW, CALIFORNIA, (415) 962-5011, TWX: 910-379-6435

MOTOROLA Semiconductors
BOX 20912 • PHOENIX, ARIZONA 85036

1N4001 thru 1N4007

Designers'Data Sheet

LEAD MOUNTED SILICON RECTIFIERS
50-1000 VOLTS
DIFFUSED JUNCTION

"SURMETIC"▲ RECTIFIERS

. . . subminiature size, axial lead mounted rectifiers for general-purpose low-power applications.

Designers Data for "Worst Case" Conditions

The Designers▲ Data Sheets permit the design of most circuits entirely from the information presented. Limit curves — representing boundaries on device characteristics — are given to facilitate "worst case" design.

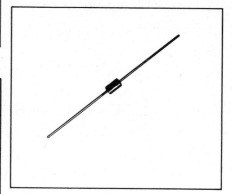

*MAXIMUM RATINGS

Rating	Symbol	1N4001	1N4002	1N4003	1N4004	1N4005	1N4006	1N4007	Unit
Peak Repetitive Reverse Voltage Working Peak Reverse Voltage DC Blocking Voltage	V_{RRM} V_{RWM} V_R	50	100	200	400	600	800	1000	Volts
Non-Repetitive Peak Reverse Voltage (halfwave, single phase, 60 Hz)	V_{RSM}	60	120	240	480	720	1000	1200	Volts
RMS Reverse Voltage	$V_{R(RMS)}$	35	70	140	280	420	560	700	Volts
Average Rectified Forward Current (single phase, resistive load, 60 Hz, see Figure 8, T_A = 75°C)	I_O				1.0				Amp
Non-Repetitive Peak Surge Current (surge applied at rated load conditions, see Figure 2)	I_{FSM}				30 (for 1 cycle)				Amp
Operating and Storage Junction Temperature Range	T_J, T_{stg}				−65 to +175				°C

*ELECTRICAL CHARACTERISTICS

Characteristic and Conditions	Symbol	Typ	Max	Unit
Maximum Instantaneous Forward Voltage Drop (i_F = 1.0 Amp, T_J = 25°C) Figure 1	v_F	0.93	1.1	Volts
Maximum Full-Cycle Average Forward Voltage Drop (I_O = 1.0 Amp, T_L = 75°C, 1 inch leads	$V_{F(AV)}$	—	0.8	Volts
Maximum Reverse Current (rated dc voltage) T_J = 25°C T_J = 100°C	I_R	0.05 1.0	10 50	μA
Maximum Full-Cycle Average Reverse Current (I_O = 1.0 Amp, T_L = 75°C, 1 inch leads	$I_{R(AV)}$	—	30	μA

*Indicates JEDEC Registered Data.

MECHANICAL CHARACTERISTICS

CASE: Void free, Transfer Molded
MAXIMUM LEAD TEMPERATURE FOR SOLDERING PURPOSES: 350°C, 3/8" from case for 10 seconds at 5 lbs. tension
FINISH: All external surfaces are corrosion-resistant, leads are readily solderable
POLARITY: Cathode indicated by color band
WEIGHT: 0.40 Grams (approximately)

▲Trademark of Motorola Inc.

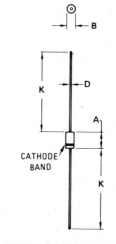

CATHODE BAND

	MILLIMETERS		INCHES	
DIM	MIN	MAX	MIN	MAX
A	5.97	6.60	0.235	0.260
B	2.79	3.05	0.110	0.120
D	0.76	0.86	0.030	0.034
K	27.94	—	1.100	—

CASE 59-04
Does Not Conform to DO-41 Outline.

© MOTOROLA INC., 1975   DS 6015 R3

MOTOROLA
Semiconductors
BOX 20912 • PHOENIX, ARIZONA 85036

2N3903
2N3904

NPN SILICON ANNULAR♦ TRANSISTORS

. . . designed for general purpose switching and amplifier applications and for complementary circuitry with types 2N3905 and 2N3906.

- High Voltage Ratings — BV_{CEO} = 40 Volts (Min)
- Current Gain Specified from 100 μA to 100 mA
- Complete Switching and Amplifier Specifications
- Low Capacitance — C_{ob} = 4.0 pF (Max)

NPN SILICON SWITCHING & AMPLIFIER TRANSISTORS

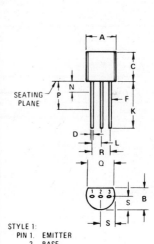

MAXIMUM RATINGS

Rating	Symbol	Value	Unit
*Collector-Base Voltage	V_{CB}	60	Vdc
*Collector-Emitter Voltage	V_{CEO}	40	Vdc
*Emitter-Base Voltage	V_{EB}	6.0	Vdc
*Collector Current	I_C	200	mAdc
Total Power Dissipation @ T_A = 60°C	P_D	250	mW
**Total Power Dissipation @ T_A = 25°C Derate above 25°C	P_D	350 2.8	mW mW/°C
**Total Power Dissipation @ T_C = 25°C Derate above 25°C	P_D	1.0 8.0	Watts mW/°C
**Junction Operating Temperature	T_J	150	°C
**Storage Temperature Range	T_{stg}	−55 to +150	°C

THERMAL CHARACTERISTICS

Characteristic	Symbol	Max	Unit
Thermal Resistance, Junction to Ambient	$R_{\theta JA}$	357	°C/W
Thermal Resistance, Junction to Case	$R_{\theta JC}$	125	°C/W

*Indicates JEDEC Registered Data
**Motorola guarantees this data in addition to the JEDEC Registered Data.
♦Annular Semiconductors Patented by Motorola Inc.

STYLE 1:
PIN 1. EMITTER
2. BASE
3. COLLECTOR

	MILLIMETERS		INCHES	
DIM	MIN	MAX	MIN	MAX
A	4.450	5.200	0.175	0.205
B	3.180	4.190	0.125	0.165
C	4.320	5.330	0.170	0.210
D	0.407	0.533	0.016	0.021
F	0.407	0.482	0.016	0.019
K	12.700	—	0.500	—
L	1.150	1.390	0.045	0.055
N	—	1.270	—	0.050
P	6.350	—	0.250	—
Q	3.430	—	0.135	—
R	2.410	2.670	0.095	0.105
S	2.030	2.670	0.080	0.105

CASE 29-02
TO-92

© MOTOROLA INC., 1973   DS 5127 R2

Characteristic		Fig. No.	Symbol	Min	Max	Unit
OFF CHARACTERISTICS						
Collector-Base Breakdown Voltage (I_C = 10 µAdc, I_E = 0)			BV$_{CBO}$	60	-	Vdc
Collector-Emitter Breakdown Voltage (1) (I_C = 1.0 mAdc, I_B = 0)			BV$_{CEO}$	40	-	Vdc
Emitter-Base Breakdown Voltage (I_E = 10 µAdc, I_C = 0)			BV$_{EBO}$	6.0	-	Vdc
Collector Cutoff Current (V_{CE} = 30 Vdc, $V_{EB(off)}$ = 3.0 Vdc)			I$_{CEX}$	-	50	nAdc
Base Cutoff Current (V_{CE} = 30 Vdc, $V_{EB(off)}$ = 3.0 Vdc)			I$_{BL}$	-	50	nAdc
ON CHARACTERISTICS						
DC Current Gain (1) (I_C = 0.1 mAdc, V_{CE} = 1.0 Vdc)	2N3903 2N3904	15	h$_{FE}$	20 40	- -	-
(I_C = 1.0 mAdc, V_{CE} = 1.0 Vdc)	2N3903 2N3904			35 70	- -	
(I_C = 10 mAdc, V_{CE} = 1.0 Vdc)	2N3903 2N3904			50 100	150 300	
(I_C = 50 mAdc, V_{CE} = 1.0 Vdc)	2N3903 2N3904			30 60	- -	
(I_C = 100 mAdc, V_{CE} = 1.0 Vdc)	2N3903 2N3904			15 30	- -	
Collector-Emitter Saturation Voltage (1) (I_C = 10 mAdc, I_B = 1.0 mAdc) (I_C = 50 mAdc, I_B = 5.0 mAdc)		16, 17	V$_{CE(sat)}$	- -	0.2 0.3	Vdc
Base-Emitter Saturation Voltage (1) (I_C = 10 mAdc, I_B = 1.0 mAdc) (I_C = 50 mAdc, I_B = 5.0 mAdc)		17	V$_{BE(sat)}$	0.65 -	0.85 0.95	Vdc
SMALL-SIGNAL CHARACTERISTICS						
Current-Gain—Bandwidth Product (I_C = 10 mAdc, V_{CE} = 20 Vdc, f = 100 MHz)	2N3903 2N3904		f$_T$	250 300	- -	MHz
Output Capacitance (V_{CB} = 5.0 Vdc, I_E = 0, f = 100 kHz)		3	C$_{ob}$	-	4.0	pF
Input Capacitance (V_{BE} = 0.5 Vdc, I_C = 0, f = 100 kHz)		3	C$_{ib}$	-	8.0	pF
Input Impedance (I_C = 1.0 mAdc, V_{CE} = 10 Vdc, f = 1.0 kHz)	2N3903 2N3904	13	h$_{ie}$	0.5 1.0	8.0 10	k ohms
Voltage Feedback Ratio (I_C = 1.0 mAdc, V_{CE} = 10 Vdc, f = 1.0 kHz)	2N3903 2N3904	14	h$_{re}$	0.1 0.5	5.0 8.0	X 10-4
Small-Signal Current Gain (I_C = 1.0 mAdc, V_{CE} = 10 Vdc, f = 1.0 kHz)	2N3903 2N3904	11	h$_{fe}$	50 100	200 400	-
Output Admittance (I_C = 1.0 mAdc, V_{CE} = 10 Vdc, f = 1.0 kHz)		12	h$_{oe}$	1.0	40	µmhos
Noise Figure (I_C = 100 µAdc, V_{CE} = 5.0 Vdc, R_S = 1.0 k ohms, f = 10 Hz to 15.7 kHz)	2N3903 2N3904	9, 10	NF	- -	6.0 5.0	dB
SWITCHING CHARACTERISTICS						
Delay Time (V_{CC} = 3.0 Vdc, $V_{BE(off)}$ = 0.5 Vdc, I_C = 10 mAdc, I_{B1} = 1.0 mAdc)		1, 5	t$_d$	-	35	ns
Rise Time		1, 5, 6	t$_r$	-	35	ns
Storage Time (V_{CC} = 3.0 Vdc, I_C = 10 mAdc, I_{B1} = I_{B2} = 1.0 mAdc)	2N3903 2N3904	2, 7	t$_s$	- -	175 200	ns
Fall Time		2, 8	t$_f$	-	50	ns

(1) Pulse Test: Pulse Width = 300µs, Duty Cycle = 2.0%.
*Indicates JEDEC Registered Data

FIGURE 1 – DELAY AND RISE TIME EQUIVALENT TEST CIRCUIT

FIGURE 2 – STORAGE AND FALL TIME EQUIVALENT TEST CIRCUIT

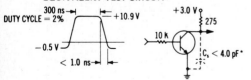

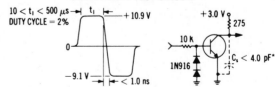

*Total shunt capacitance of test jig and connectors

Ⓜ **MOTOROLA** *Semiconductor Products Inc.*

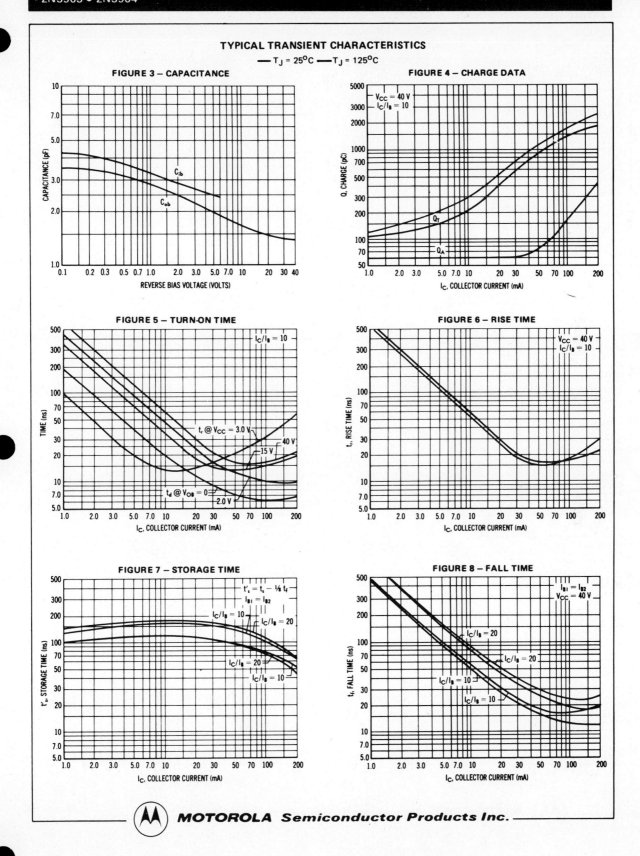

TYPICAL TRANSIENT CHARACTERISTICS

— T_J = 25°C — T_J = 125°C

FIGURE 3 – CAPACITANCE

FIGURE 4 – CHARGE DATA

FIGURE 5 – TURN-ON TIME

FIGURE 6 – RISE TIME

FIGURE 7 – STORAGE TIME

FIGURE 8 – FALL TIME

MOTOROLA *Semiconductor Products Inc.*

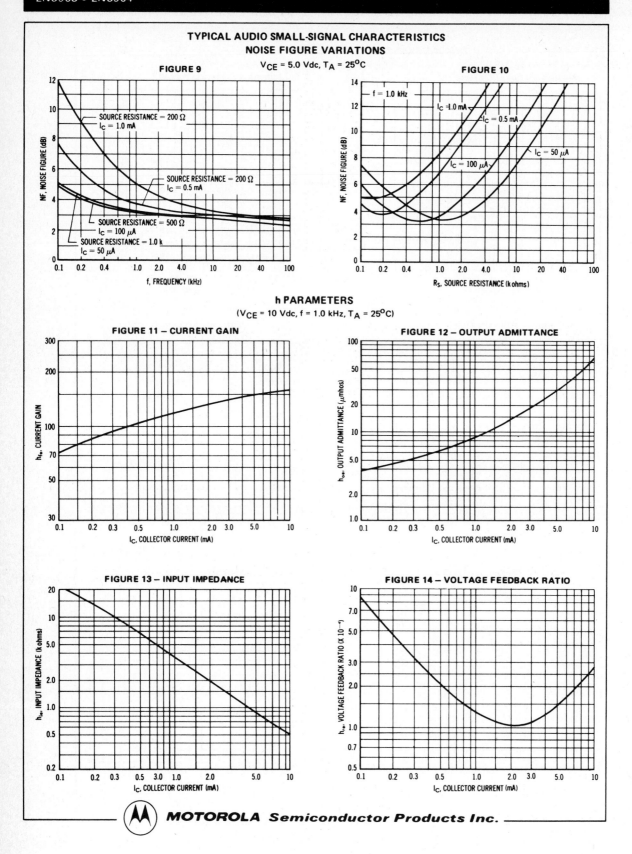

TYPICAL AUDIO SMALL-SIGNAL CHARACTERISTICS
NOISE FIGURE VARIATIONS
V_{CE} = 5.0 Vdc, T_A = 25°C

FIGURE 9

FIGURE 10

h PARAMETERS
(V_{CE} = 10 Vdc, f = 1.0 kHz, T_A = 25°C)

FIGURE 11 — CURRENT GAIN

FIGURE 12 — OUTPUT ADMITTANCE

FIGURE 13 — INPUT IMPEDANCE

FIGURE 14 — VOLTAGE FEEDBACK RATIO

MOTOROLA *Semiconductor Products Inc.*

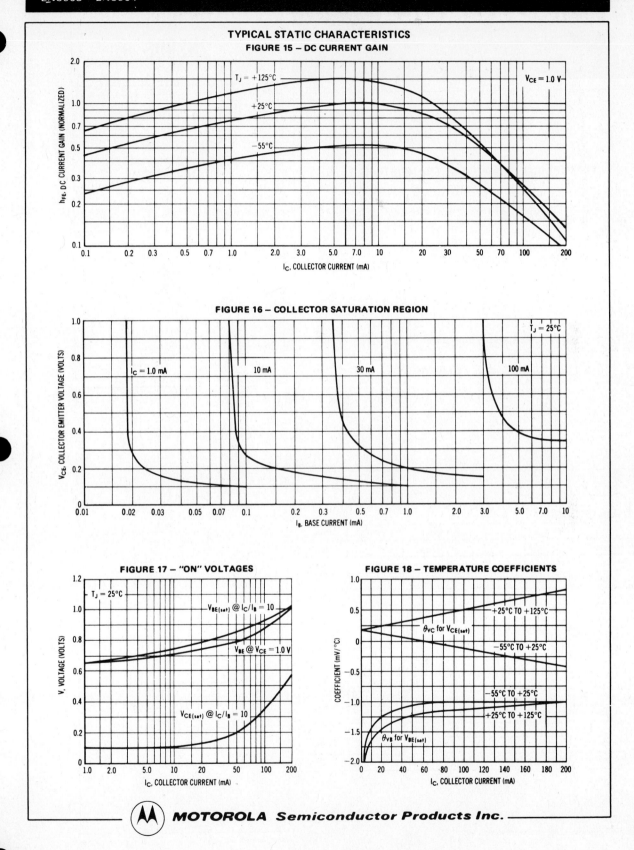

TYPICAL STATIC CHARACTERISTICS
FIGURE 15 – DC CURRENT GAIN

FIGURE 16 – COLLECTOR SATURATION REGION

FIGURE 17 – "ON" VOLTAGES

FIGURE 18 – TEMPERATURE COEFFICIENTS

MOTOROLA *Semiconductor Products Inc.*

MOTOROLA
Semiconductors
BOX 20912 • PHOENIX. ARIZONA 85036

2N3905
2N3906

PNP SILICON ANNULAR♦ TRANSISTORS

.... designed for general purpose switching and amplifier applications and for complementary circuitry with types 2N3903 and 2N3904.

- High Voltage Ratings — BV_{CEO} = 40 Volts (Min)
- Current Gain Specified from 100 μA to 100 mA
- Complete Switching and Amplifier Specifications
- Low Capacitance — C_{ob} = 4.5 pF (Max)

**PNP SILICON
SWITCHING & AMPLIFIER
TRANSISTORS**

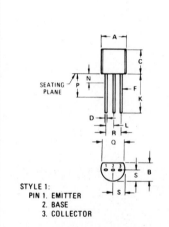

STYLE 1:
PIN 1. EMITTER
2. BASE
3. COLLECTOR

DIM	MILLIMETERS		INCHES	
	MIN	MAX	MIN	MAX
A	4.450	5.200	0.175	0.205
B	3.180	4.190	0.125	0.165
C	4.320	5.330	0.170	0.210
D	0.407	0.533	0.016	0.021
F	0.407	0.482	0.016	0.019
K	12.700	—	0.500	—
L	1.150	1.390	0.045	0.055
N	—	1.270	—	0.050
P	6.350	—	0.250	—
Q	3.430	—	0.135	—
R	2.410	2.670	0.095	0.105
S	2.030	2.670	0.080	0.105

CASE 29-02
(TO-92)

*MAXIMUM RATINGS

Rating	Symbol	Value	Unit
Collector-Base Voltage	V_{CB}	40	Vdc
Collector-Emitter Voltage	V_{CEO}	40	Vdc
Emitter-Base Voltage	V_{EB}	5.0	Vdc
Collector Current	I_C	200	mAdc
Total Power Dissipation @ $T_A = 60^oC$	P_D	250	mW
Total Power Dissipation @ $T_A = 25^oC$ Derate above 25^oC	P_D	350 2.8	mW mW/^oC
Total Power Dissipation @ $T_C = 25^oC$ Derate above 25^oC	P_D	1.0 8.0	Watt mW/^oC
Junction Operating Temperature	T_J	+150	^oC
Storage Temperature Range	T_{stg}	−55 to +150	^oC

THERMAL CHARACTERISTICS

Characteristic	Symbol	Max	Unit
Thermal Resistance, Junction to Ambient	$R_{\theta JA}$	357	^oC/W
Thermal Resistance, Junction to Case	$R_{\theta JC}$	125	^oC/W

*Indicates JEDEC Registered Data.
♦Annular semiconductors patented by Motorola Inc.

DS 5128 R2

*ELECTRICAL CHARACTERISTICS ($T_A = 25°C$ unless otherwise noted.)

Characteristic		Fig. No.	Symbol	Min	Max	Unit
OFF CHARACTERISTICS						
Collector-Base Breakdown Voltage ($I_C = 10\,\mu Adc$, $I_E = 0$)			BV_{CBO}	40	—	Vdc
Collector-Emitter Breakdown Voltage (1) ($I_C = 1.0\,mAdc$, $I_B = 0$)			BV_{CEO}	40	—	Vdc
Emitter-Base Breakdown Voltage ($I_E = 10\,\mu Adc$, $I_C = 0$)			BV_{EBO}	5.0	—	Vdc
Collector Cutoff Current ($V_{CE} = 30\,Vdc$, $V_{BE(off)} = 3.0\,Vdc$)			I_{CEX}	—	50	nAdc
Base Cutoff Current ($V_{CE} = 30\,Vdc$, $V_{BE(off)} = 3.0\,Vdc$)			I_{BL}	—	50	nAdc
ON CHARACTERISTICS (1)						
DC Current Gain			h_{FE}			
($I_C = 0.1\,mAdc$, $V_{CE} = 1.0\,Vdc$)	2N3905	15		30	—	
	2N3906			60	—	
($I_C = 1.0\,mAdc$, $V_{CE} = 1.0\,Vdc$)	2N3905			40	—	
	2N3906			80	—	
($I_C = 10\,mAdc$, $V_{CE} = 1.0\,Vdc$)	2N3905			50	150	
	2N3906			100	300	
($I_C = 50\,mAdc$, $V_{CE} = 1.0\,Vdc$)	2N3905			30	—	
	2N3906			60	—	
($I_C = 100\,mAdc$, $V_{CE} = 1.0\,Vdc$)	2N3905			15	—	
	2N3906			30	—	
Collector-Emitter Saturation Voltage		16, 17	$V_{CE(sat)}$			Vdc
($I_C = 10\,mAdc$, $I_B = 1.0\,mAdc$)				—	0.25	
($I_C = 50\,mAdc$, $I_B = 5.0\,mAdc$)				—	0.4	
Base-Emitter Saturation Voltage		17	$V_{BE(sat)}$			Vdc
($I_C = 10\,mAdc$, $I_B = 1.0\,mAdc$)				0.65	0.85	
($I_C = 50\,mAdc$, $I_B = 5.0\,mAdc$)				—	0.95	
SMALL-SIGNAL CHARACTERISTICS						
Current-Gain — Bandwidth Product			f_T			MHz
($I_C = 10\,mAdc$, $V_{CE} = 20\,Vdc$, $f = 100\,MHz$)	2N3905			200	—	
	2N3906			250	—	
Output Capacitance ($V_{CB} = 5.0\,Vdc$, $I_E = 0$, $f = 100\,kHz$)		3	C_{ob}	—	4.5	pF
Input Capacitance ($V_{BE} = 0.5\,Vdc$, $I_C = 0$, $f = 100\,kHz$)		3	C_{ib}	—	1.0	pF
Input Impedance		13	h_{ie}			k ohms
($I_C = 1.0\,mAdc$, $V_{CE} = 10\,Vdc$, $f = 1.0\,kHz$)	2N3906			0.5	8.0	
	2N3906			2.0	12	
Voltage Feedback Ratio		14	h_{re}			X 10⁻⁴
($I_C = 1.0\,mAdc$, $V_{CE} = 10\,Vdc$, $f = 1.0\,kHz$)	2N3905			0.1	5.0	
	2N3906			1.0	10	
Small-Signal Current Gain		11	h_{fe}			—
($I_C = 1.0\,mAdc$, $V_{CE} = 10\,Vdc$, $f = 1.0\,kHz$)	2N3905			50	200	
	2N3906			100	400	
Output Admittance		12	h_{oe}			μmhos
($I_C = 1.0\,mAdc$, $V_{CE} = 10\,Vdc$, $f = 1.0\,kHz$)	2N3905			1.0	40	
	2N3906			3.0	60	
Noise Figure		9, 10	NF			dB
($I_C = 100\,\mu Adc$, $V_{CE} = 5.0\,Vdc$, $R_S = 1.0\,k\,ohm$,	2N3905			—	5.0	
$f = 10\,Hz$ to $15.7\,kHz$)	2N3906			—	4.0	
SWITCHING CHARACTERISTICS						
Delay Time ($V_{CC} = 3.0\,Vdc$, $V_{BE(off)} = 0.5\,Vdc$		1, 5	t_d	—	35	ns
Rise Time $\quad I_C = 10\,mAdc$, $I_{B1} = 1.0\,mAdc$)		1, 5, 6	t_r	—	35	ns
Storage Time	2N3905	2, 7	t_s	—	200	ns
	2N3906			—	225	
Fall Time ($V_{CC} = 3.0\,Vdc$, $I_C = 10\,mAdc$,	2N3905	2, 8	t_f	—	60	ns
$I_{B1} = I_{B2} = 1.0\,mAdc$)	2N3906			—	75	

*Indicates JEDEC Registered Data. (1) Pulse Width = 300 μs, Duty Cycle = 2.0 %.

FIGURE 1 – DELAY AND RISE TIME EQUIVALENT TEST CIRCUIT

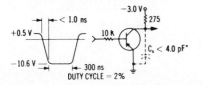

FIGURE 2 – STORAGE AND FALL TIME EQUIVALENT TEST CIRCUIT

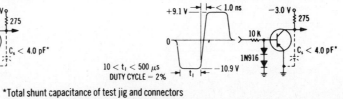

*Total shunt capacitance of test jig and connectors

TRANSIENT CHARACTERISTICS
— $T_J = 25°C$ — $T_J = 125°C$

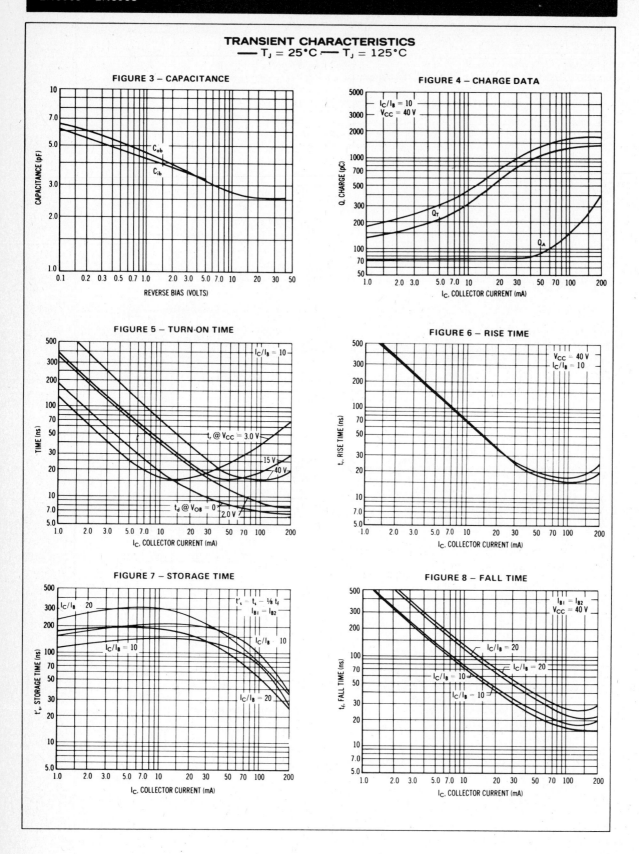

FIGURE 3 – CAPACITANCE

FIGURE 4 – CHARGE DATA

FIGURE 5 – TURN-ON TIME

FIGURE 6 – RISE TIME

FIGURE 7 – STORAGE TIME

FIGURE 8 – FALL TIME

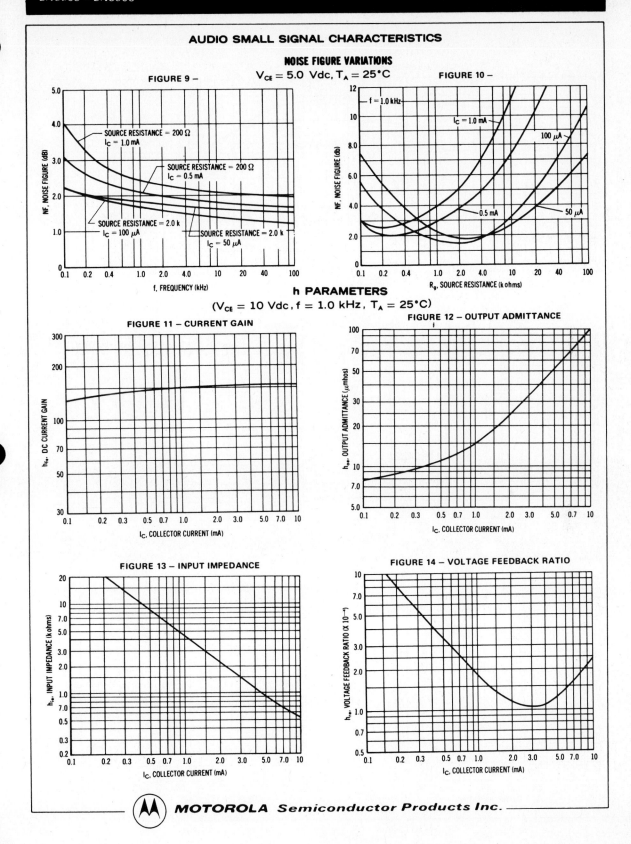

AUDIO SMALL SIGNAL CHARACTERISTICS

NOISE FIGURE VARIATIONS
$V_{CE} = 5.0$ Vdc, $T_A = 25°C$

FIGURE 9 –

FIGURE 10 –

h PARAMETERS
($V_{CE} = 10$ Vdc, $f = 1.0$ kHz, $T_A = 25°C$)

FIGURE 11 – CURRENT GAIN

FIGURE 12 – OUTPUT ADMITTANCE

FIGURE 13 – INPUT IMPEDANCE

FIGURE 14 – VOLTAGE FEEDBACK RATIO

MOTOROLA *Semiconductor Products Inc.*

STATIC CHARACTERISTICS

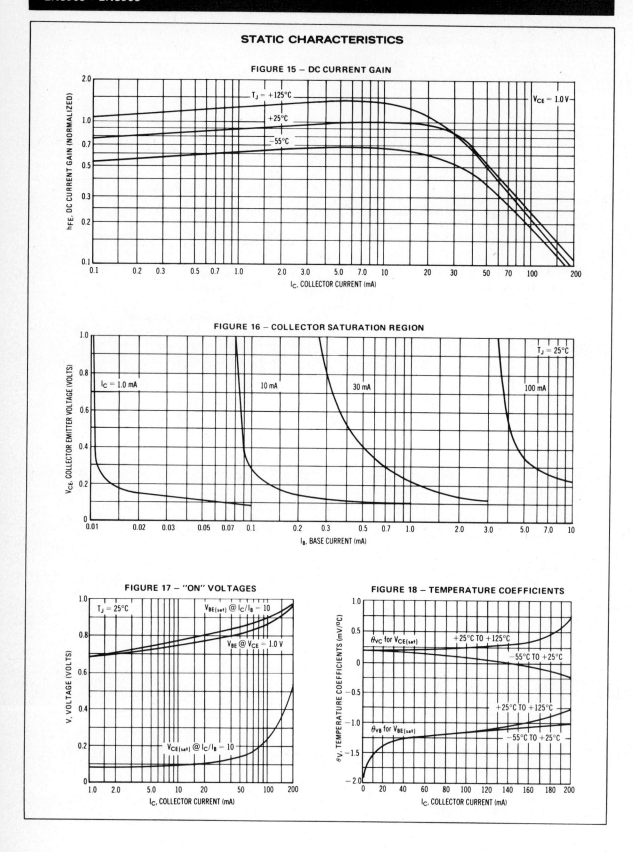

FIGURE 15 – DC CURRENT GAIN

FIGURE 16 – COLLECTOR SATURATION REGION

FIGURE 17 – "ON" VOLTAGES

FIGURE 18 – TEMPERATURE COEFFICIENTS

MOTOROLA Semiconductors
BOX 20912 • PHOENIX, ARIZONA 85036

2N4441 thru 2N4444

PLASTIC SILICON CONTROLLED RECTIFIERS

8.0 AMPERES RMS
50 thru 600 VOLTS

PLASTIC THYRISTORS

. . . designed for high-volume consumer phase-control applications such as motor speed, temperature, and light controls and for switching applications in ignition and starting systems, voltage regulators, vending machines, and lamp drivers requiring:

- Small, Rugged, Thermopad ▲ Construction — for Low Thermal Resistance, High Heat Dissipation, and Durability.

- Practical Level Triggering and Holding Characteristics @ 25°C
  I_{GT} = 7.0 mA (Typ)
  I_H = 6.0 mA (Typ)

- Low "On" Voltage – V_{TM} = 1.0 Volt (Typ) @ 5.0 Amp @ 25°C

- High Surge Current Rating – I_{TSM} = 80 Amp

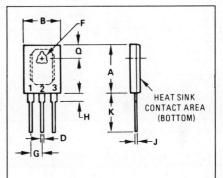

MAXIMUM RATING (T_J = 100°C unless otherwise noted.)

Rating	Symbol	Value	Unit
*Repetitive Peak Reverse Blocking Voltage (Note 1)	V_{RRM}		Volts
2N4441		50	
2N4442		200	
2N4443		400	
2N4444		600	
*Non-Repetitive Peak Reverse Blocking Voltage (t = 5.0 ms (max) duration)	V_{RSM}		Volts
2N4441		75	
2N4442		300	
2N4443		500	
2N4444		700	
*RMS On-State Current (All Conduction Angles)	$I_{T(RMS)}$	8.0	Amp
Average On-State Current, T_C = 73°C	$I_{T(AV)}$	5.1	Amp
*Peak Non-Repetitive Surge Current (1/2 cycle, 60 Hz preceded and followed by rated current and voltage)	I_{TSM}	80	Amp
Circuit Fusing Considerations (T_J = –40 to +100°C; t = 1.0 to 8.3 ms)	I^2t	25	A^2s
*Peak Gate Power	P_{GM}	5.0	Watts
*Average Gate Power	$P_{G(AV)}$	0.5	Watt
*Peak Forward Gate Current	I_{GM}	2.0	Amp
*Peak Reverse Gate Voltage	V_{RGM}	10	Volts
*Operating Junction Temperature Range	T_J	–40 to +100	°C
*Storage Temperature Range	T_{stg}	–40 to +150	°C
Mounting Torque (6-32 screw) (Note 2)	—	8.0	in. lb.

THERMAL CHARACTERISTICS

Characteristic	Symbol	Typ	Max	Unit
*Thermal Resistance, Junction to Case	$R_{\theta JC}$	—	2.5	°C/W
Thermal Resistance, Junction to Ambient	$R_{\theta JA}$	40	—	°C/W

*Indicates JEDEC Registered Data.
▲Trademark of Motorola Inc.

DIM	MILLIMETERS		INCHES	
	MIN	MAX	MIN	MAX
A	15.95	16.71	0.628	0.658
B	12.45	13.21	0.490	0.520
C	3.05	3.81	0.120	0.150
D	1.09	1.25	0.043	0.049
F	3.51	3.76	0.138	0.148
G	4.22 BSC		0.166 BSC	
H	—	3.18	—	0.125
J	0.76	0.86	0.030	0.034
K	14.99	16.51	0.590	0.650
Q	4.50	5.00	0.177	0.197
R	1.91	2.16	0.075	0.085

CASE 90-04

STYLE 1:
PIN 1. CATHODE
2. ANODE
3. GATE

HEAT SINK CONTACT AREA (BOTTOM)

© MOTOROLA INC., 1973   DS 6533 R1

ELECTRICAL CHARACTERISTICS (T_C = 25°C unless otherwise noted)

Characteristic	Symbol	Min	Typ	Max	Unit

*Peak Forward Blocking Voltage
(T_J = 100°C) Note 1   2N4441
    2N4442
    2N4443
    2N4444
	V_{DRM}	50
		200
		400
		600
			—
			—
			—
			—
				—
				—
				—
				—
					Volts

Peak Forward Blocking Current (Rated V_{DRM}, T_J = 100°C, gate open)	I_{DRM}	—	—	2.0	mA
Peak Reverse Blocking Current (Rated V_{DRM}, T_J = 100°C, gate open)	I_{RRM}	—	—	2.0	mA
Gate Trigger Current (Continuous dc) (Anode Voltage = 7.0 Vdc, R_L = 100 Ohms) T_C = 25°C * T_C = –40°C	I_{GT}	— 	7.0 	30 60	mA

Gate Trigger Voltage (Continuous dc)
(Anode Voltage = 7.0 Vdc, R_L = 100 Ohms)
    T_C = 25°C
* (Anode Voltage = 7.0 Vdc, R_L = 100 Ohms)
    T_C = –40°C
* (Anode Voltage = Rated V_{DRM}, R_L = 100 Ohms)
    T_J = 100°C
	V_{GT}	—
		
		—
		
		0.2
			0.75
			
			—
			
			—
				1.5
				
				2.5
				
				—
					Volts

Peak On-State Voltage (Pulse Width = 1.0 to 2.0 ms, Duty Cycle ⩽ 2.0%) (I_{TM} = 5.0 A peak) * (I_{TM} = 15.7 A peak)	V_{TM}	— —	1.0 —	1.5 2.0	Volts
Holding Current (Anode Voltage = 7.0 Vdc, gate open) T_C = 25°C * T_C = –40°C	I_H	— —	6.0 —	40 70	mA
Gate Controlled Turn-On Time (I_{TM} = 5.0 A, I_{GT} = 20 mA)	t_{gt}	—	1.0	—	µs
Circuit Commutated Turn-Off Time (I_{TM} = 5.0 A, I_R = 5.0 A) (I_{TM} = 5.0 A, I_R = 5.0 A, T_J = 100°C)	t_q	— —	15 20	— —	µs
Critical Rate of Rise of Off-State Voltage (Rated V_{DRM}, Exponential Waveform, T_J = 100°C, Gate Open)	dv/dt	—	50	—	V/µs

*Indicates JEDEC Registered Data

Note 1.   Ratings apply for zero or negative gate voltage but positive gate voltage shall not be applied concurrently with a negative potential on the anode. When checking forward or reverse blocking capability, thyristor devices should not be tested with a constant current source in a manner that the voltage applied exceeds the rated blocking voltage.

Note 2.   Torque rating applies with use of torque washer (Shakeproof WD19522 #6 or equivalent). Mounting torque in excess of 8 in. lbs. does not appreciably lower case-to-sink thermal resistance. Anode lead and heatsink contact pad are common.

For soldering purposes (either terminal connection or device mounting), soldering temperatures shall not exceed +225°C.

MOTOROLA *Semiconductor Products Inc.*

MOTOROLA

**JUNCTION
FIELD-EFFECT
TRANSISTOR**

**SYMMETRICAL
SILICON
N-CHANNEL**

SEPTEMBER 1966 — DS 5203

SILICON N-CHANNEL
JUNCTION FIELD-EFFECT TRANSISTOR

. . . designed for VHF amplifier and mixer applications.

- Low Cross-Modulation and Intermodulation Distortion
- Guaranteed 100-MHz Parameters
- Drain and Source Interchangeable
- Low Transfer and Input Capacitance
- Low Leakage Current
- Unibloc* Plastic Encapsulated Package

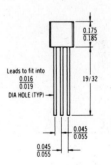

"D" shape package lies flat for easy printed circuit mounting.

Rugged, one-piece, high-temperature, pressure-molded, humidity resistant, plastic package

19/32 inch, gold-plated nickel, oval leads permit reliable solder connections.

MAXIMUM RATINGS (T_A 25°C)

Characteristic	Symbol	Rating	Unit
Drain-Source Voltage	V_{DS}	25	Vdc
Drain-Gate Voltage	V_{DG}	25	Vdc
Gate-Source Voltage	V_{GS}	–25	Vdc
Gate Current	I_G	10	mAdc
Total Device Dissipation Derate above 25°C	P_D	200 2	mW mW/°C
Operating Junction Temperature	T_J	125	°C
Storage Temperature Range	T_{stg}	–65 to +150	°C

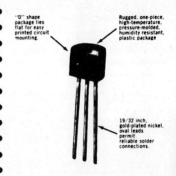

Leads to fit into
0.016
0.019
DIA HOLE (TYP)

0.175
0.185

19/32

0.045
0.055

0.045
0.055

SOURCE
0.003
0.013 R.

5° (TYP)

DRAIN

GATE

0.085
0.095 R.

0.045
0.055

BOTTOM VIEW

TO-92

Drain and Source may be Interchanged.

°Trademark of Motorola Inc.

MOTOROLA Semiconductor Products Inc.  A SUBSIDIARY OF MOTOROLA INC.

SI FIELD-EFFECT TRANSISTOR
MPF102
DS 5203

ELECTRICAL CHARACTERISTICS (T$_A$ = 25°C unless otherwise noted)

Characteristic	Symbol	Min	Max	Unit
OFF CHARACTERISTICS				
Gate-Source Breakdown Voltage ($I_G = -10 \mu$Adc, $V_{DS} = 0$)	BV$_{GSS}$	-25	—	Vdc
Gate Reverse Current ($V_{GS} = -15$ Vdc, $V_{DS} = 0$)	I$_{GSS}$	—	-2.0	nAdc
($V_{GS} = -15$ Vdc, $V_{DS} = 0$, $T_A = 100$°C)		—	-2.0	μAdc
Gate-Source Cutoff Voltage ($V_{DS} = 15$ Vdc, $I_D = 2.0$ nAdc)	V$_{GS(off)}$	—	-8	Vdc
Gate-Source Voltage ($V_{DS} = 15$ Vdc, $I_D = 0.2$ mAdc)	V$_{GS}$	-0.5	-7.5	Vdc
ON CHARACTERISTICS				
Zero-Gate-Voltage Drain Current* ($V_{DS} = 15$ Vdc, $V_{GS} = 0$ Vdc)	I$_{DSS}$*	2	20	mAdc
DYNAMIC CHARACTERISTICS				
Forward Transfer Admittance* ($V_{DS} = 15$ Vdc, $V_{GS} = 0$, $f = 1$ kHz)	$\|y_{fs}\|$*	2000	7500	μmhos
Input Capacitance ($V_{DS} = 15$ Vdc, $V_{GS} = 0$, $f = 1$ MHz)	C$_{iss}$	—	7	pF
Reverse Transfer Capacitance ($V_{DS} = 15$ Vdc, $V_{GS} = 0$, $f = 1$ MHz)	C$_{rss}$	—	3	pF
Forward Transfer Admittance ($V_{DS} = 15$ Vdc, $V_{GS} = 0$, $f = 100$ MHz)	$\|y_{fs}\|$	1600	—	μmhos
Input Conductance ($V_{DS} = 15$ Vdc, $V_{GS} = 0$, $f = 100$ MHz)	Re(y$_{is}$)	—	800	μmhos
Output Conductance ($V_{DS} = 15$ Vdc, $V_{GS} = 0$, $f = 100$ MHz)	Re(y$_{os}$)	—	200	μmhos

*Pulse Test: Pulse Width ≤ 630 ms; Duty Cycle ≤ 10%

MOTOROLA *Semiconductor Products Inc.*

BOX 955 • PHOENIX. ARIZONA 85001 • A SUBSIDIARY OF MOTOROLA INC.

1813-1 PRINTED IN USA 3-67 IMPERIAL LITHO B1467

DS 5203

National Semiconductor

LM118/LM218/LM318 operational amplifier

general description

The LM118 series are precision high speed operational amplifiers designed for applications requiring wide bandwidth and high slew rate. They feature a factor of ten increase in speed over general purpose devices without sacrificing DC performance.

features

- 15 MHz small signal bandwidth
- Guaranteed 50V/μs slew rate
- Maximum bias current of 250 nA
- Operates from supplies of ±5V to ±20V
- Internal frequency compensation
- Input and output overload protected
- Pin compatible with general purpose op amps

The LM118 series has internal unity gain frequency compensation. This considerably simplifies its application since no external components are necessary for operation. However, unlike most internally

compensated amplifiers, external frequency compensation may be added for optimum performance For inverting applications, feedforward compensation will boost the slew rate to over 150V/μs and almost double the bandwidth. Overcompensation can be used with the amplifier for greater stability when maximum bandwidth is not needed. Further, a single capacitor can be added to reduce the 0.1% settling time to under 1 μs.

The high speed and fast settling time of these op amps make them useful in A/D converters, oscillators, active filters, sample and hold circuits, or general purpose amplifiers. These devices are easy to apply and offer an order of magnitude better AC performance than industry standards such as the LM709.

The LM218 is identical to the LM118 except that the LM218 has its performance specified over a −25°C to +85°C temperature range. The LM318 is specified from 0°C to +70°C.

schematic and connection diagrams

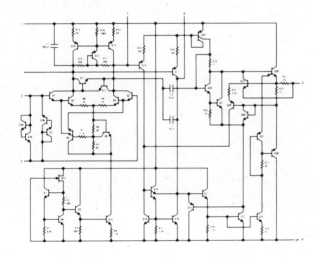

Flat Package

Order Number LM118F or LM218F
See Package 3

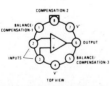

Metal Can Package*

*Pin connections shown on schematic diagram and typical applications are for TO-5 package.

Order Number LM118H, LM218H
or LM318H
See Package 11

Dual-In-Line Package

Order Number LM318N
See Package 20

Dual-In-Line Package

Order Number LM118D, LM218D
or LM318D
See Package 1

3-135

absolute maximum ratings

Supply Voltage	±20V
Power Dissipation (Note 1)	500 mW
Differential Input Current (Note 2)	±10 mA
Input Voltage (Note 3)	±15V
Output Short-Circuit Duration	Indefinite
Operating Temperature Range	
LM118	−55°C to +125°C
LM218	−25°C to +85°C
LM318	0°C to +70°C
Storage Temperature Range	−65°C to +150°C
Lead Temperature (Soldering, 10 seconds)	300°C

electrical characteristics (Note 4)

PARAMETER	CONDITIONS	LM118/LM218 MIN	TYP	MAX	LM318 MIN	TYP	MAX	UNITS
Input Offset Voltage	T_A = 25°C		2	4		4	10	mV
Input Offset Current	T_A = 25°C		6	50		30	200	nA
Input Bias Current	T_A = 25°C		120	250		150	500	nA
Input Resistance	T_A = 25°C	1	3		0.5	3		MΩ
Supply Current	T_A = 25°C		5	8		5	10	mA
Large Signal Voltage Gain	T_A = 25°C, V_S = ±15V V_{OUT} = ±10V, $R_L \geq$ 2 kΩ	50	200		25	200		V/mV
Slew Rate	T_A = 25°C, V_S = ±15V, A_V = 1	50	70		50	70		V/µs
Small Signal Bandwidth	T_A = 25°C, V_S = ±15V		15			15		MHz
Input Offset Voltage				6			15	mV
Input Offset Current				100			300	nA
Input Bias Current				500			750	nA
Supply Current	T_A = 125°C		4.5	7				
Large Signal Voltage Gain	V_S = ±15V, V_{OUT} = ±10V $R_L \geq$ 2 kΩ	25			20			V/mV
Output Voltage Swing	V_S = ±15V, R_L = 2 kΩ	±12	±13		±12	±13		V
Input Voltage Range	V_S = ±15V	±11.5			±11.5			V
Common-Mode Rejection Ratio		80	100		70	100		dB
Supply Voltage Rejection Ratio		70	80		65	80		dB

Note 1: The maximum junction temperature of the LM118 is 150°C, the LM218 is 110°C, and the LM318 is 110°C. For operating at elevated temperatures, devices in the TO-5 package must be derated based on a thermal resistance of 150°C/W, junction to ambient, or 45°C/W, junction to case. For the flat package, the derating is based on a thermal resistance of 185°C/W when mounted on a 1/16-inch-thick epoxy glass board with ten, 0.03-inch-wide, 2-ounce copper conductors. The thermal resistance of the dual-in-line package is 100°C/W, junction to ambient.

Note 2: The inputs are shunted with back-to-back diodes for overvoltage protection. Therefore, excessive current will flow if a differential input voltage in excess of 1V is applied between the inputs unless some limiting resistance is used.

Note 3: For supply voltages less than ±15V, the absolute maximum input voltage is equal to the supply voltage.

Note 4: These specifications apply for ±5V $\leq V_S \leq$ ±20V and −55°C $\leq T_A \leq$ +125°C, (LM118), −25°C $\leq T_A \leq$ +85°C (LM218), and 0°C $\leq T_A \leq$ +70°C (LM318). Also, power supplies must be bypassed with 0.1µF disc capacitors.

3-136

typical performance characteristics LM318

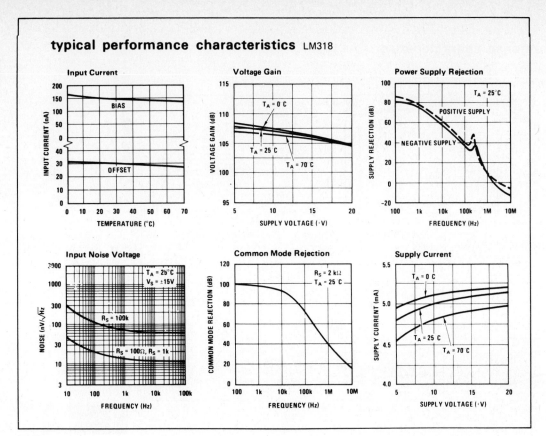

3-138

typical performance characteristics LM318 (Cont'd)

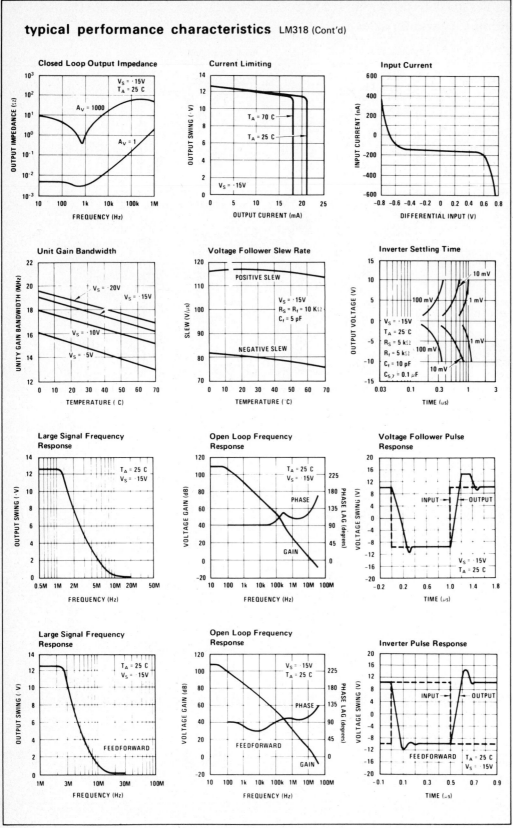

3-139

LM140A/LM140/LM340A/LM340 Series
3-Terminal Positive Regulators

General Description

The LM140A/LM140/LM340A/LM340 series of positive 3-terminal voltage regulators are designed to provide superior performance as compared to the previously available 78XX series regulator. Computer programs were used to optimize the electrical and thermal performance of the packaged IC which results in outstanding ripple rejection, superior line and load regulation in high power applications (over 15W).

With these advances in design, the LM340 is now guaranteed to have line and load regulation that is a factor of 2 better than previously available devices. Also, all parameters are guaranteed at 1A vs 0.5A output current. The LM140A/LM340A provide tighter output voltage tolerance, ±2% along with 0.01%/V line regulation and 0.3%/A load regulation.

Current limiting is included to limit peak output current to a safe value. Safe area protection for the output transistor is provided to limit internal power dissipation. If internal power dissipation becomes too high for the heat sinking provided, the thermal shutdown circuit takes over limiting die temperature.

Considerable effort was expended to make the LM140-XX series of regulators easy to use and minimize the number of external components. It is not necessary to bypass the output, although this does improve transient response.

Input bypassing is needed only if the regulator is located far from the filter capacitor of the power supply.

Although designed primarily as fixed voltage regulators, these devices can be used with external components to obtain adjustable voltages and currents.

The entire LM140A/LM140/LM340A/LM340 series of regulators is available in the metal TO-3 power package and the LM340A/LM340 series is also available in the TO-220 plastic power package.

Features

- Complete specifications at 1A load
- Output voltage tolerances of ±2% at $T_j = 25°C$ and ±4% over the temperature range (LM140A/LM340A)
- Fixed output voltages available 5, 6, 8, 10 12, 15, 18 and 24V
- Line regulation of 0.01% of V_{OUT}/V ΔV_{IN} at 1A load (LM140A/LM340A)
- Load regulation of 0.3% of V_{OUT}/A ΔI_{LOAD} (LM140A/LM340A)
- Internal thermal overload protection
- Internal short-circuit current limit
- Output transistor safe area protection

Typical Applications

Fixed Output Regulator

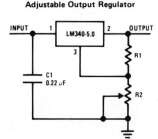

*Required if the regulator is located far from the power supply filter

**Although no output capacitor is needed for stability, it does help transient response. (If needed, use 0.1 μF, ceramic disc)

Adjustable Output Regulator

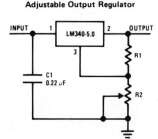

$V_{OUT} = 5V + (5V/R1 + I_Q) R2$

$5V/R1 > 3 I_Q$, load regulation $(L_r) \approx$
$[(R1 + R2)/R1]$ $(L_r$ of LM340-5)

Current Regulator

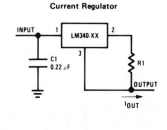

$I_{OUT} = \dfrac{V_{2-3}}{R1} + I_Q$

$\Delta I_Q = 1.3$ mA over line and load changes

Electrical Characteristics LM340 (Note 2)

$0°C \leq T_j \leq +125°C$ unless otherwise noted.

Column groups are OUTPUT VOLTAGE / INPUT VOLTAGE: 5V/10V, 6V/11V, 8V/14V, 10V/17V, 12V/19V, 15V/23V, 18V/27V, 24V/33V. Each has MIN, TYP, MAX.

Parameter	Conditions	5V MIN	5V TYP	5V MAX	6V MIN	6V TYP	6V MAX	8V MIN	8V TYP	8V MAX	10V MIN	10V TYP	10V MAX	12V MIN	12V TYP	12V MAX	15V MIN	15V TYP	15V MAX	18V MIN	18V TYP	18V MAX	24V MIN	24V TYP	24V MAX	Units
V_O Output Voltage	$T_j = 25°C$, $5\,mA \leq I_O \leq 1A$	4.8	5	5.2	5.75	6	6.25	7.7	8	8.3	9.6	10	10.4	11.5	12	12.5	14.4	15	15.6	17.3	18	18.7	23.0	24	25.0	V
	$P_D \leq 15W$, $5\,mA \leq I_O \leq 1A$	4.75		5.25	5.7		6.3	7.6		8.4	9.5		10.5	11.4		12.6	14.25		15.75	17.1		18.9	22.8		25.2	V
	$V_{MIN} \leq V_{IN} \leq V_{MAX}$		$(7 \leq V_{IN} \leq 20)$			$(8 \leq V_{IN} \leq 21)$			$(10.5 \leq V_{IN} \leq 23)$			$(12.5 \leq V_{IN} \leq 25)$			$(14.5 \leq V_{IN} \leq 27)$			$(17.5 \leq V_{IN} \leq 30)$			$(21 \leq V_{IN} \leq 33)$			$(27 \leq V_{IN} \leq 38)$		V
ΔV_O Line Regulation	$I_O \leq 500\,mA$, $T_j = 25°C$		3	50		3	60		4	80		4	100		4	120		4	150		4	180		6	240	mV
	ΔV_{IN}		$(7 \leq V_{IN} \leq 25)$			$(8 \leq V_{IN} \leq 25)$			$(10.5 \leq V_{IN} \leq 25)$			$(12.5 \leq V_{IN} \leq 25)$			$(14.5 \leq V_{IN} \leq 27)$			$(17.5 \leq V_{IN} \leq 30)$			$(21 \leq V_{IN} \leq 33)$			$(27 \leq V_{IN} \leq 38)$		V
	$0°C \leq T_j \leq +125°C$			50			60			80			100			120			150			180			240	mV
	ΔV_{IN}		$(8 \leq V_{IN} \leq 20)$			$(9 \leq V_{IN} \leq 21)$			$(11 \leq V_{IN} \leq 23)$			$(13 \leq V_{IN} \leq 25)$			$(15 \leq V_{IN} \leq 27)$			$(18.5 \leq V_{IN} \leq 30)$			$(21.5 \leq V_{IN} \leq 33)$			$(28 \leq V_{IN} \leq 38)$		V
	$I_O \leq 1A$, $T_j = 25°C$			25			30			40			50			60			75			90			120	mV
	ΔV_{IN}		$(7.3 \leq V_{IN} \leq 20)$			$(8.35 \leq V_{IN} \leq 21)$			$(10.5 \leq V_{IN} \leq 23)$			$(12.5 \leq V_{IN} \leq 25)$			$(14.6 \leq V_{IN} \leq 27)$			$(17.7 \leq V_{IN} \leq 30)$			$(21 \leq V_{IN} \leq 33)$			$(27.1 \leq V_{IN} \leq 38)$		V
	$0°C \leq T_j \leq +125°C$			25			30			40			50			60			75			90			120	mV
	ΔV_{IN}		$(8 \leq V_{IN} \leq 12)$			$(9 \leq V_{IN} \leq 13)$			$(11 \leq V_{IN} \leq 17)$			$(14 \leq V_{IN} \leq 20)$			$(16 \leq V_{IN} \leq 22)$			$(20 \leq V_{IN} \leq 26)$			$(24 \leq V_{IN} \leq 30)$			$(30 \leq V_{IN} \leq 36)$		V
ΔV_O Load Regulation	$T_j = 25°C$, $5\,mA \leq I_O \leq 1.5A$		10	50		12	60		12	80		12	100		12	120		12	150		12	180		12	240	mV
	$250\,mA \leq I_O \leq 750\,mA$			25			30			40			50			60			75			90			120	mV
	$5\,mA \leq I_O \leq 1A$, $0°C \leq T_j \leq +125°C$			50			60			80			100			120			150			180			240	mV
I_Q Quiescent Current	$I_O \leq 1A$, $T_j = 25°C$		8	8.5		8	8.5		8	8.5		8	8.5		8	8.5		8	8.5		8	8.5		8	8.5	mA
	$0°C \leq T_j \leq +125°C$			8.5			8.5			8.5			8.5			8.5			8.5			8.5			8.5	mA
ΔI_Q Quiescent Current Change	$5\,mA \leq I_O \leq 1A$			0.5			0.5			0.5			0.5			0.5			0.5			0.5			0.5	mA
	$T_j = 25°C$, $I_O \leq 1A$, $V_{MIN} \leq V_{IN} \leq V_{MAX}$			1.0			1.0			1.0			1.0			1.0			1.0			1.0			1.0	mA
			$(7.5 \leq V_{IN} \leq 20)$			$(8.6 \leq V_{IN} \leq 21)$			$(10.6 \leq V_{IN} \leq 23)$			$(12.7 \leq V_{IN} \leq 25)$			$(14.8 \leq V_{IN} \leq 27)$			$(17.9 \leq V_{IN} \leq 30)$			$(21 \leq V_{IN} \leq 33)$			$(27.3 \leq V_{IN} \leq 38)$		V
	$I_O \leq 500\,mA$, $0°C \leq T_j \leq +125°C$, $V_{MIN} \leq V_{IN} \leq V_{MAX}$			1.0			1.0			1.0			1.0			1.0			1.0			1.0			1.0	mA
			$(7 \leq V_{IN} \leq 25)$			$(8 \leq V_{IN} \leq 25)$			$(10.5 \leq V_{IN} \leq 25)$			$(12.5 \leq V_{IN} \leq 25)$			$(14.5 \leq V_{IN} \leq 30)$			$(17.5 \leq V_{IN} \leq 30)$			$(21 \leq V_{IN} \leq 38)$			$(27 \leq V_{IN} \leq 38)$		V
V_N Output Noise Voltage	$T_A = 25°C$, $10\,Hz \leq f \leq 100\,kHz$		40			45			52			70			75			90			110			170		µV
$\dfrac{\Delta V_{IN}}{\Delta V_{OUT}}$ Ripple Rejection	$f = 120\,Hz$ $\{\ I_O = 1A, T_j = 25°C\ or$	62			59			56			55			55			54			53			50			dB
	$I_O \leq 500\,mA$, $0°C \leq T_j \leq +125°C\ \}$	62	80		59	78		56	76		55	74		55	72		54	70		53	69		50	66		dB
	$V_{MIN} \leq V_{IN} \leq V_{MAX}$		$(8 \leq V_{IN} \leq 18)$			$(9 \leq V_{IN} \leq 19)$			$(11.5 \leq V_{IN} \leq 21.5)$			$(13.5 \leq V_{IN} \leq 23.5)$			$(15 \leq V_{IN} \leq 25)$			$(18.5 \leq V_{IN} \leq 28.5)$			$(22 \leq V_{IN} \leq 32)$			$(28 \leq V_{IN} \leq 38)$		V
R_O Dropout Voltage	$T_j = 25°C$, $I_{OUT} = 1A$		2.0			2.0			2.0			2.0			2.0			2.0			2.0			2.0		V
Output Resistance	$f = 1\,kHz$		8			9			12			16			18			19			22			28		mΩ
Short Circuit Current	$T_j = 25°C$		2.1			2.0			1.9			1.7			1.5			1.2			0.8			0.4		A
Peak Output Current	$T_j = 25°C$		2.4			2.4			2.4			2.4			2.4			2.4			2.4			2.4		A
Average TC of V_{OUT}	$0°C \leq T_j \leq +150°C$, $I_O = 5\,mA$		-0.6			0.7			1.0			1.2			1.5			-1.8			-2.3			-3.0		mV/°C
V_{IN} Input Voltage Required to Maintain Line Regulation	$T_j = 25°C$, $I_O \leq 1A$		7.3			8.35			10.5			12.5			14.6			17.7			21			27.1		V

Note 2: All characteristics are measured with a capacitor across the input of 0.22 µF and a capacitor across the output of 0.1 µF. All characteristics except noise voltage and ripple rejection ratio are measured using pulse techniques ($t_w \leq 10\,ms$, duty cycle $\leq 5\%$). Output voltage changes due to changes in internal temperature must be taken into account separately.

Typical Performance Characteristics

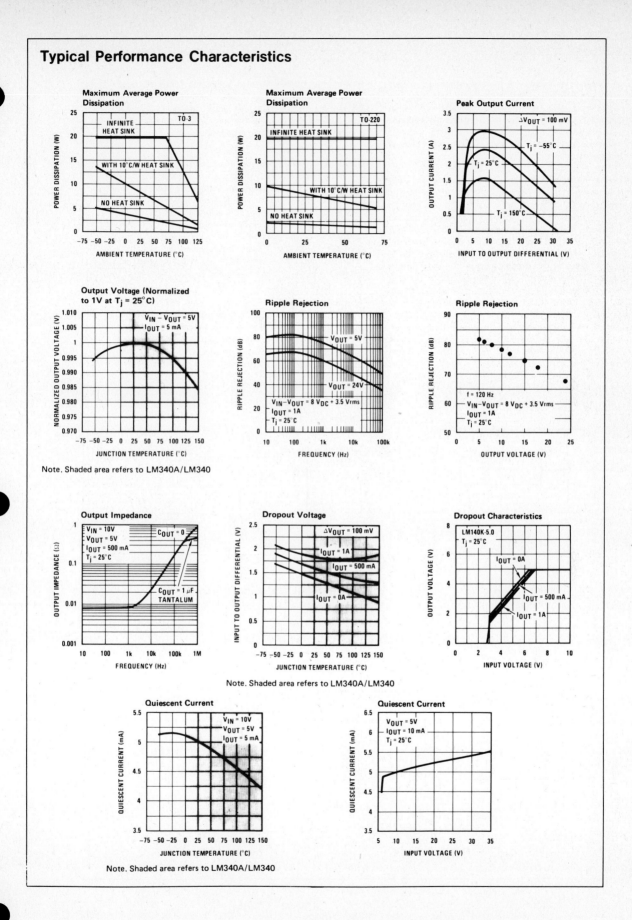

Note. Shaded area refers to LM340A/LM340

Note. Shaded area refers to LM340A/LM340

Note. Shaded area refers to LM340A/LM340

Application Hints

The LM340 is designed with thermal protection, output short-circuit protection and output transistor safe area protection. However, as with *any* IC regulator, it becomes necessary to take precautions to assure that the regulator is not inadvertently damaged. The following describes possible misapplications and methods to prevent damage to the regulator.

Shorting the Regulator Input: When using large capacitors at the output of these regulators that have V_{OUT} greater than 6V, a protection diode connected input to output *(Figure 1)* may be required if the input is shorted to ground. Without the protection diode, an input short will cause the input to rapidly approach ground potential, while the output remains near the initial V_{OUT} because of the stored charge in the large output capacitor. The capacitor will then discharge through reverse biased emitter-base junction of the pass device, Q16, which breaks down at 6.5V and forward biases the base-collector junction. If the energy released by the capacitor into the emitter-base junction is large enough, the junction and the regulator will be destroyed. The fast diode in *Figure 1* will shunt the capacitor's discharge current around the regulator.

Raising the Output Voltage above the Input Voltage: Since the output of the LM340 does not sink current, forcing the output high can cause damage to internal low current paths in a manner similar to that just described in the "Shorting the Regulator Input" section.

Regulator Floating Ground *(Figure 2)*: When the ground pin alone becomes disconnected, the output approaches the unregulated input, causing possible damage to other circuits connected to V_{OUT}. If ground is reconnected with power "ON", damage may also occur to the regulator. This fault is most likely to occur when plugging in regulators or modules with on card regulators into powered up sockets. Power should be turned off first, or ground should be connected first if power must be left on.

Transient Voltages: If transients exceed the maximum rated input voltage of the 340, or reach more than 0.8V below ground and have sufficient energy, they will damage the regulator. The solution is to use a large input capacitor, a series input breakdown diode, a choke, a transient suppressor or a combination of these.

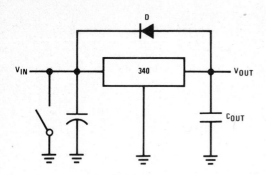

FIGURE 1. Input Short

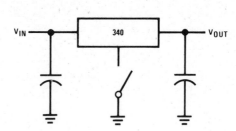

FIGURE 2. Regulator Floating Ground

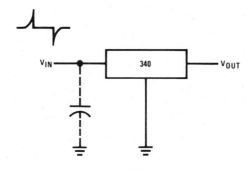

FIGURE 3. Transients

Connection Diagrams

Metal Can Package

BOTTOM VIEW

Pin 1 — input
Pin 2 — output
Case — ground

Power Package

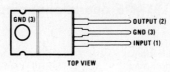

TOP VIEW

LM741/LM741A/LM741C/LM741E operational amplifier

general description

The LM741 series are general purpose operational amplifiers which feature improved performance over industry standards like the LM709. They are direct, plug-in replacements for the 709C, LM201, MC1439 and 748 in most applications.

The amplifiers offer many features which make their application nearly foolproof: overload protection on the input and output, no latch-up when the common mode range is exceeded, as well as freedom from oscillations.

The LM741C/LM741E are identical to the LM741/LM741A except that the LM741C/LM741E have their performance guaranteed over a 0°C to +70°C temperature range, instead of −55°C to +125°C.

schematic and connection diagrams (Top Views)

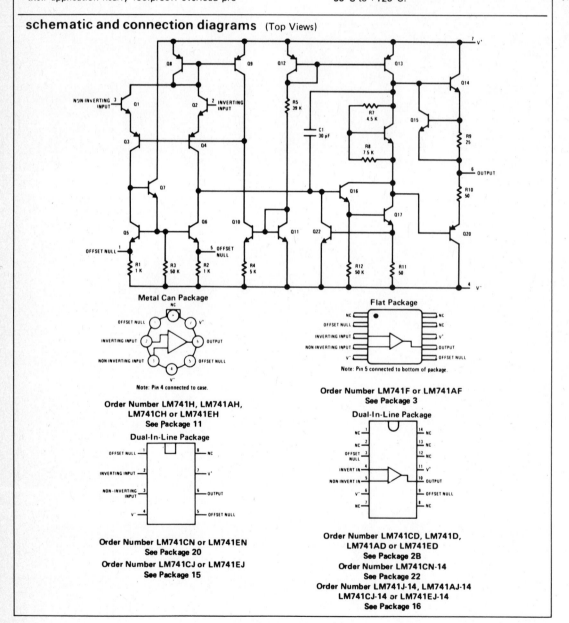

Metal Can Package

Note: Pin 4 connected to case.

Order Number LM741H, LM741AH, LM741CH or LM741EH
See Package 11

Dual-In-Line Package

Order Number LM741CN or LM741EN
See Package 20
Order Number LM741CJ or LM741EJ
See Package 15

Flat Package

Note: Pin 5 connected to bottom of package.

Order Number LM741F or LM741AF
See Package 3

Dual-In-Line Package

Order Number LM741CD, LM741D, LM741AD or LM741ED
See Package 2B
Order Number LM741CN-14
See Package 22
Order Number LM741J-14, LM741AJ-14 LM741CJ-14 or LM741EJ-14
See Package 16

LM741/LM741A/LM741C/LM741E

absolute maximum ratings

	LM741A	LM741E	LM741	LM741C
Supply Voltage	±22V	±22V	±22V	±18V
Power Dissipation (Note 1)	500 mW	500 mW	500 mW	500 mW
Differential Input Voltage	±30V	±30V	±30V	±30V
Input Voltage (Note 2)	±15V	±15V	±15V	±15V
Output Short Circuit Duration	Indefinite	Indefinite	Indefinite	Indefinite
Operating Temperature Range	−55°C to +125°C	0°C to +70°C	−55°C to +125°C	0°C to +70°C
Storage Temperature Range	−65°C to +150°C	−65°C to +150°C	−65°C to +150°C	−65°C to +150°C
Lead Temperature (Soldering, 10 seconds)	300°C	300°C	300°C	300°C

electrical characteristics (Note 3)

PARAMETER	CONDITIONS	LM741A/LM741E MIN	TYP	MAX	LM741 MIN	TYP	MAX	LM741C MIN	TYP	MAX	UNITS
Input Offset Voltage	$T_A = 25°C$										
	$R_S \leq 10\ k\Omega$					1.0	5.0		2.0	6.0	mV
	$R_S \leq 50\Omega$		0.8	3.0							mV
	$T_{AMIN} \leq T_A \leq T_{AMAX}$										
	$R_S \leq 50\Omega$			4.0							mV
	$R_S \leq 10\ k\Omega$						6.0			7.5	mV
Average Input Offset Voltage Drift				15							µV/°C
Input Offset Voltage Adjustment Range	$T_A = 25°C$, $V_S = ±20V$	±10				±15			±15		mV
Input Offset Current	$T_A = 25°C$		3.0	30		20	200		20	200	nA
	$T_{AMIN} \leq T_A < T_{AMAX}$			70		85	500			300	nA
Average Input Offset Current Drift				0.5							nA/°C
Input Bias Current	$T_A = 25°C$		30	80		80	500		80	500	nA
	$T_{AMIN} \leq T_A \leq T_{AMAX}$			0.210			1.5			0.8	µA
Input Resistance	$T_A = 25°C$, $V_S = ±20V$	1.0	6.0		0.3	2.0		0.3	2.0		MΩ
	$T_{AMIN} \leq T_A \leq T_{AMAX}$, $V_S = ±20V$	0.5									MΩ
Input Voltage Range	$T_A = 25°C$							±12	±13		V
	$T_{AMIN} \leq T_A \leq T_{AMAX}$				±12	±13					V
Large Signal Voltage Gain	$T_A = 25°C$, $R_L \geq 2\ k\Omega$										
	$V_S = ±20V$, $V_O = ±15V$	50									V/mV
	$V_S = ±15V$, $V_O = ±10V$				50	200		20	200		V/mV
	$T_{AMIN} \leq T_A \leq T_{AMAX}$, $R_L \geq 2\ k\Omega$,										
	$V_S = ±20V$, $V_O = ±15V$	32									V/mV
	$V_S = ±15V$, $V_O = ±10V$				25			15			V/mV
	$V_S = ±5V$, $V_O = ±2V$	10									V/mV
Output Voltage Swing	$V_S = ±20V$										
	$R_L \geq 10\ k\Omega$	±16									V
	$R_L \geq 2\ k\Omega$	±15									V
	$V_S = ±15V$										
	$R_L \geq 10\ k\Omega$				±12	±14		±12	±14		V
	$R_L \geq 2\ k\Omega$				±10	±13		±10	±13		V
Output Short Circuit Current	$T_A = 25°C$	10	25	35		25			25		mA
	$T_{AMIN} < T_A < T_{AMAX}$	10		40							mA
Common-Mode Rejection Ratio	$T_{AMIN} \leq T_A \leq T_{AMAX}$										
	$R_S \leq 10\ k\Omega$, $V_{CM} = ±12V$				70	90		70	90		dB
	$R_S \leq 50\ k\Omega$, $V_{CM} = ±12V$	80	95								dB

electrical characteristics (con't)

PARAMETER	CONDITIONS	LM741A/LM741E			LM741			LM741C			UNITS
		MIN	TYP	MAX	MIN	TYP	MAX	MIN	TYP	MAX	
Supply Voltage Rejection Ratio	$T_{AMIN} \le T_A \le T_{AMAX}$, $V_S = \pm20V$ to $V_S = \pm5V$										
	$R_S \le 50\Omega$	86	96								dB
	$R_S \le 10 \, k\Omega$				77	96		77	96		dB
Transient Response	$T_A = 25°C$, Unity Gain										
Rise Time			0.25	0.8		0.3			0.3		μs
Overshoot			6.0	20		5			5		%
Bandwidth (Note 4)	$T_A = 25°C$	0.437	1.5								MHz
Slew Rate	$T_A = 25°C$, Unity Gain	0.3	0.7			0.5			0.5		$V/\mu s$
Supply Current	$T_A = 25°C$					1.7	2.8		1.7	2.8	mA
Power Consumption	$T_A = 25°C$										
	$V_S = \pm20V$		80	150							mW
	$V_S = \pm15V$					50	85		50	85	mW
LM741A	$V_S = \pm20V$										
	$T_A = T_{AMIN}$			165							mW
	$T_A = T_{AMAX}$			135							mW
LM741E	$V_S = \pm20V$			150							mW
	$T_A = T_{AMIN}$			150							mW
	$T_A = T_{AMAX}$			150							mW
LM741	$V_S = \pm15V$										
	$T_A = T_{AMIN}$					60	100				mW
	$T_A = T_{AMAX}$					45	75				mW

Note 1: The maximum junction temperature of the LM741/LM741A is 150°C, while that of the LM741C/LM741E is 100°C. For operation at elevated temperatures, devices in the TO-5 package must be derated based on a thermal resistance of 150°C/W junction to ambient, or 45°C/W junction to case. The thermal resistance of the dual-in-line package is 100°C/W junction to ambient. For the flat package, the derating is based on a thermal resistance of 185°C/W when mounted on a 1/16 inch thick epoxy glass board with ten, 0.03 inch wide, 2 ounce copper conductors.

Note 2: For supply voltages less than ±15V, the absolute maximum input voltage is equal to the supply voltage.

Note 3: Unless otherwise specified, these specifications apply for $V_S = \pm15V$, $-55°C \le T_A \le +125°C$ (LM741/LM741A). For the LM741C/LM741E, these specifications are limited to $0°C \le T_A \le +70°C$.

Note 4: Calculated value from: BW (MHz) = 0.35/Rise Time(μs).